Powdered Crude Drug Microscopy of Leaves and Barks

Powdered Crude Drug Microscopy of Leaves and Barks

Vidhu Aeri

D B Anantha Narayana

Dharya Singh

ELSEVIER

Elsevier
Radarweg 29, PO Box 211, 1000 AE Amsterdam, Netherlands
The Boulevard, Langford Lane, Kidlington, Oxford OX5 1GB, United Kingdom
50 Hampshire Street, 5th Floor, Cambridge, MA 02139, United States

Notices

Knowledge and best practice in this field are constantly changing. As new research and experience broaden our understanding, changes in research methods, professional practices, or medical treatment may become necessary.

Practitioners and researchers must always rely on their own experience and knowledge in evaluating and using any information, methods, compounds, or experiments described herein. In using such information or methods they should be mindful of their own safety and the safety of others, including parties for whom they have a professional responsibility.

To the fullest extent of the law, neither the Publisher nor the authors, contributors, or editors, assume any liability for any injury and/or damage to persons or property as a matter of products liability, negligence or otherwise, or from any use or operation of any methods, products, instructions, or ideas contained in the material herein.

Library of Congress Cataloging-in-Publication Data
A catalog record for this book is available from the Library of Congress

British Library Cataloguing-in-Publication Data
A catalogue record for this book is available from the British Library

ISBN: 978-0-12-818092-1

For information on all Elsevier publications visit our website at
https://www.elsevier.com/books-and-journals

Publisher: Susan Dennis
Acquisition Editor: Emil M. McCloskey
Editorial Project Manager: Lindsay Lawrence
Production Project Manager: Omer Mukthar
Cover Designer: Matthew Limbert

Typeset by TNQ Technologies

Contents

Biography

Vidhu Aeri, PhD

Dr. Vidhu Aeri obtained her Master's degree from Delhi University and her PhD from Jamia Hamdard, New Delhi. Presently, she is working as a Professor, in the Department of Pharmacognosy & Phytochemistry, School of Pharmaceutical Education and Research, Jamia Hamdard, New Delhi. She is Involved in teaching and research since 1989.

For the last 30 years, Vidhu has been working on medicinal plants with reference to their pharmacognostical, phytochemical, analytical and pharmacological work leading to quality control and substantiating their traditional and documented medicinal uses and claims.

She has advised 14 doctoral and 60 postgraduate students. She has completed projects from ICMR, UGC, DST, NMPB, Govt. of India. She has worked as a Production Manager, Bakson Homeo Pharmacy Pvt. Ltd., New Delhi before taking up academics.

She has authored three books and six book chapters and more than 100 research papers in National and International Journals to her credit.

Presently, Dr. Aeri is Head of the Department and Subject Expert Member, to evaluate new phytopharmaceutical drugs proposals for regulatory approval, Central Drugs Standard Control Organization (CDSCO), Directorate General of Health Services, MOHFW, Govt. of India.

She is also Chairman, Subcommittee, Guidance document, SOP preparation/ amendment and Review, Phytopharmaceutical Division, Indian Pharmacopoeia Commission (IPC), Govt. of India.

D. B. Anantha Narayana, PhD

Dr. Narayana, a recipient of the Eminent Pharmacist Award 2007 of the Indian Pharmaceutical Association, is an M Pharm from Saugar University (MP), PhD from Delhi University. He taught pharmacy for ten years and worked at Ranbaxy Laboratories for over ten years.

He later joined Dabur Research Foundation (DRF) and led a team of researchers to develop and launch anti-cancer drugs, ayurvedic drugs, cosmetics, food products, honey and many other herbal products. At Dabur and later at HUL, he helped the demonstration of the application of "Reverse Pharmacology" approach to develop herbal products with contemporary scientific validations.

He served in HUL/Unilever Research, for nine years and was instrumental in creating a strong regulatory affairs group for South Asia. During his tenure, Brook Bond Natural Care Tea was proven clinically to help build immunity.

He is Currently Chairman of Phytopharmaceuticals and Herbals Committee of Indian Pharmacopoeia Commission (IPC). He is an Expert member of the Committees of Food Safety & Standards Authority of India (FSSAI).

He founded AYURVIDYE Research & Training Trust to promote Ayurveda, in 2010. Blending science with regulatory aspects he championed the emergence of two important regulations recently notified - for supplements and nutraceuticals in 2016/17 under Indian Foods Regulations, as well as phytopharmaceuticals as drugs under Indian Drug Laws in 2016. He chaired the Herbs and Herbal Products Committee of Indian Pharmacopoeia Commission for over ten years. He led development and inclusion of Quality monographs of over 150 herbs in Indian Pharmacopoeia so far.

He was an expert member in a scientific panel on nutraceuticals of the Food Safety Standards Authority of India for eight years. Dr. Narayana has received a number of awards for his contributions to science and profession, namely the Acharya P. C. Ray Gold Medal award, Dr. K. M. Parikh Memorial award, Dr. Govinda Achari award, Sibte Hasan Zaidi oration award, K. L. E. University oration award, Indian Drugs Outstanding Contribution award, the Patanjali Life time Achievement award of IUFoST 2018.

Dr. Dharya Singh, PhD

Dr. Dharya Singh obtained her PhD in the thesis entitled 'Computer Aided Drug Microscopy of Some Leafy and Bark Drugs', Department of Pharmacognosy and Phytochemistry, Jamia Hamdard, New Delhi in the year 2018. The research work was aimed at developing a software/databank TIPHAM for identification of powdered samples of leaves and barks used in Indian Traditional Systems of Medicine along with their substitutes and adulterants.

She qualified GPAT 2010 with 99.18 percentile (AIR-266) and worked as a part time lecturer in Delhi Institute of Pharmaceutical Sciences and Research (DIPSAR), Pushp Vihar, New Delhi for three years (2013–2016). She has published six research papers in reputed journals and participated in national and international conferences.

Preface

Herbs or botanicals have been used as dietary supplements, nutraceuticals, food additives, ingredients in cosmetics and flavours and fragrances in addition to their use in traditional herbal/drug preparations like Ayurvedic and TCM drugs. Industrial usage of herbs from plants and their parts, either as such or after processing has gone up in the last decade. The usage approximately comprises of 50% roots, 15% seeds, 12% wood waste, 9% whole plants, 7% barks/stems, 4% leaves and 3% flowers as raw material. However, the technology for confirming their botanical identity is still evolving. Before subjecting samples of herbs for their quality, authenticity of the genus and species, prior to industrial use is needed. Many difficulties do exist in the operation and implementation of confirming the botanical identity of parts of herb samples. The fear of substitution and adulteration of raw herbs is also on the rise as most of the time, only plant parts rather than whole drugs are available in the commercial market.

The knowledge, competency and skills for evaluation of microscopical features of herb samples need focused development. Use of microscopic techniques is not at the levels desired, though they are powerful tools to identify herbs and their powdered materials. It is a known fact that industry is moving towards procurement of powdered herbs rather than whole herbs and their subsequent use in their respective formulations/preparations. This is to avoid difficulty in grinding and associated Good Manufacturing Practices compliance due to generation of dust during the grinding process.

Performing macroscopy and microscopy of each plant part in every consignment in industry is a tedious and time-consuming job. Moreover, these procedures are often not well-documented, raising doubt about the identification of herbs, translating into ineffective quality control of the end-product.

Concurrently, scientists trained in pharmacognosy and taxonomy are few, and in such a scenario, training non-pharmacognosy graduates to perform quality control and authentication of herbs has its challenges. Pharmacopoeias have introduced mandatory thin layer chromatography (TLC) testing of plant material under examination in the last decade or so. This has improved identity testing, subject to availability of specific marker compounds, but cannot replace botanical identity testing.

The manufacture, sale and distribution of herbal products are regulated in India under the Drugs and Cosmetics Act, 1940 and Rules, 1945. Similar regulations exist in other nations for herbal preparations. WHO and other organizations emphasize the need for quality and standardization of plants used in manufacturing of traditional medicines.

In India, Ayurvedic Pharmacopoeia, Indian Pharmacopoeia, Indian Herbal Pharmacopoeia and ICMR monographs have standards for checking the authenticity of botanicals used in herbal products. However, this is easier said than done, as the sheer number of applications and the magnitude of usage of herbs have gone up, which poses a challenge due to the presence of adulterants, substitutes and controversial nomenclature.

Microscopic identification is the most commonly used method for authentication of herbal drugs. By means of using various microscopic techniques, structural and cellular features of herbs are examined in order to determine their botanical origin. This method is useful for identifying species with similar morphological characters. Today, there are a variety of methods available to authenticate herbal drugs, ranging from simple morphological examination to physical and chemical analysis, and DNA molecular biology. Owing to their cost, powder microscopy is still the most practical method for primary authentication. Many reference books, namely the Ayurvedic Pharmacopoeia of India, British Pharmacopoeia, Chinese Pharmacopoeia, European Pharmacopoeia, Japanese Pharmacopoeia, United States Pharmacopoeia and Vietnamese Pharmacopoeia, all record the powder microscopic characters of herbal drugs.

Botanical microscopy is a unique, valuable, rapid and cost-effective assessment tool. It continues to play an important role in the authentication and assessment of medicinal plants. The study of pharmacognostic properties of raw herbs and their identification appears to have not gotten the attention it deserves. In such a scenario, this book provides a fundamental understanding of the macroscopic and microscopic characters of herbs in raw form and more so in their powdered form. A total of 52 barks and leaves with adulterants and substitutes from commonly used herbs listed in different pharmacopoeias of Indian Traditional Systems are covered.

The features selected for identification of bark powder include nine general characters, namely calcium oxalate crystals, starch grains, medullary rays, cork, tubular lactiferous canals, phloem parenchyma, masses, rhytidoma, and secretory canals, and three specific characters, namely stone cells, sclereids and fibres. These specific characters are further differentiated into 12 features and 127 subfeatures.

The features selected for identification of leaf powder include 11 general characters, namely epidermis, calcium oxalate crystals, fibers, stone cells and sclereids, cell contents, cystoliths, lamina, starch grains, tracheids, lactiferous canals, and xylem vessels and two specific characters, namely stomata and trichomes, which are further differentiated into 56 features.

The book was developed, based on the specificity of these parameters for a particular bark and leaf in its powdered form. Photographs of the macroscopy and microscopical features of the crude herbs have been provided.

A chapter has been included on the grinding of plant materials, describing the guidelines for GMP during grinding, as well as discussing advanced grinding techniques like cryo-grinding, and packing of ground herbs and their storage. Standard Operating Protocols for preparation of slides to observe general as well as distinguishing microscopical characters of barks and leaves were developed by comparing samples of powdered herbs ground differently. Samples ground in the laboratory were compared with samples ground using industrial grinders. Data showed that the grinding technique does not significantly affect the microscopical features of the crude herbs.

The images given in this book are real pictorial representations of various microscopical features. Researchers can relate and understand these features more clearly during microscopic analysis. It is hoped that the contents of the book would be found useful by undergraduate, postgraduate and PhD research students involved in the study of plants and natural products as well as professionals in industries manufacturing plant-based products.

I acknowledge the continuous support of Jamia Hamdard and Ayurvidye Trust, Bengaluru, for their involvement in this work. The authors acknowledge Mr. Pradeep Aeri for his constant valuable support in photography to ensure the closest original representation of crude drugs.

Prof. (Dr.) Vidhu Aeri
New Delhi, India
29th June, 2019

Chapter 1

Introduction to powder microscopy

Herbal medicines have become an essential part of healthcare throughout the world. There are several medicinal plants that are used as medicine as single herbs in minimal processed forms or as starting material for the synthesis of chemical drugs or as models for pharmacologically active compounds. With such an enormous expansion in the use of herbal medicines globally, authentication of herbs has become an important concern for manufacturers, researchers and the general public. World Health Assembly has emphasized the need to have proper regulatory measures and the need to ensure the quality of medicinal plants. The therapeutic efficacy of herbal medicine depends on the genuineness of ingredients used in preparation of formulation.

In the historic and prehistoric times, the question of procurement of drugs and controlling their quality did not arise because physician himself used to take care of this aspect. He used to collect his own drugs, process them into formulations and prescribe them to patients. But as the time advanced, the population of patients increased, demand for medicines increased, the process of urbanization accelerated, leading to marked depletion of the natural source of herbal drugs. Thus, collection of drugs was taken over by commercial agencies and supply of drugs became a business. In many cases, botanically different plant species are sold in place of Ayurvedic, Unani and Homoeopathic drugs. Sometimes different species of the same genus that are easily and cheaply available are used in place of the drug, which has been prescribed in the literature.

The plant-based crude drugs are often adulterated/substituted or obtained from sources other than the prescribed one. It is a mistaken notion that classical procedures are now obsolete and only the chemotaxonomical profile can provide a true identity of the drug and help to distinguish them from substitutes and adulterants. Microscopic authentication refers to observing cell structure and internal features using a microscope and its derivatives. Besides the ordinary light microscope, other microscopes also have been used to enhance the accuracy of authentication, such as the polarized light microscope and fluorescence microscope. Powder microscopy used in pharmacognosy is still a useful tool for species identification and can play a vital role in

Powdered Crude Drug Microscopy of Leaves and Barks. https://doi.org/10.1016/B978-0-12-818092-1.00001-4

authentication of drug material. Microscopic characters can be standardized for a particular herbal drug by examining microscopic images.

Churna or powder consisting of a single drug or a combination of more than one drug is one of the most commonly used prescription in Ayurveda and microscopic evaluation of such drugs is the easiest method for determining their correct identity. Barks and leaves from trees and shrubs constitute an important source of drugs used in Indian System of Medicine. After collection and drying, many barks resemble each other morphologically but microscopic characters of their powders exhibit marked differences, assisting greatly in their correct botanical identity. Recent advancements in normal light microscopy have greatly enhanced its usefulness in the authentication of herbal medicines. Feature extraction and similarity measurement have been used effectively in the classification of microscopic images of powdered herbal drugs. This provides greater accuracy and flexibility in capturing information about microscopical features, which are useful in authentication of herbal drugs.

There has been a revival of interest in herbal products for treatment of variety of diseases. Herbal drugs have been used as dietary supplements, nutraceuticals, food additives, as ingredients in cosmetics and flavours and fragrances. There is a dramatic increase in popularity and acceptance of herbal medicines. The industrial usage of plants and their parts, either as such or after processing, has gone up, but the technology for confirming their botanical identity is still questionable. The fear of substitution and adulteration of raw herbs is also on rise as most of the times only plant parts are available in the industry: 50% roots, 15% seeds, 12% wood waste, 9% whole plants, 7% bark/stem, 4% leaves and 3% flowers are used as raw material. With such an enormous increase in the use of herbal drugs, it is the need of the hour to have a check on the correct botanical identity of the plant. Performing macroscopy and microscopy of each plant part in every consignment in industry is a tedious, time consuming job and one does not see good documentation leaving doubts about the quality control of the tests being done. Some of the pharmacopoeias have introduced mandatory thin-layer chromatography (TLC) testing of the plant material under examination. This has improved the identity testing, but cannot replace the botanical identity testing as proper identification of plant is the key to develop a formulation of utmost quality and safety.

The manufacture, sale and distribution of herbal products are regulated in India under Drugs and Cosmetics Act 1940 and Rules 1945. The WHO and other organizations emphasize the need for quality and standardization of plants used in manufacturing of traditional medicines including Ayurveda, where the first basic requirement is establishing the correct botanical identity of plant drug attributing it to a specific genus and species. In India Ayurvedic Pharmacopoeia, Indian Pharmacopoeia, Indian Herbal Pharmacopoeia and ICMR monographs have standards for checking authenticity of botanicals used in herbal products. However, serious efforts are not made to resolve the issue of controversial nomenclature. The long history of safe usage of herbal medicines can be extrapolated only when the botanical identity of the plant going into those medicines is established and standardized.

The aspect of identification and study of pharmacognostic properties of raw herbs appears to have not got the attention it deserves. Powdered drugs are used in herbal industries in formulations. In order to ensure that final product is safe and efficacious, its quality should be checked at first stage. Microscopic identification is the most commonly used method for authentication of herbal medicines. By means of using various microscopic techniques, structural and cellular features of herbs are examined in order to determine their botanical origins and assess their qualities. This method is useful for identifying species with similar morphological characters. Today, there are a variety of methods available to authenticate herbal drugs, ranging from simple morphological examination to physical and chemical analysis, and DNA molecular biology. Owing to their cost, powder microscopy is still the most practical method for primary authentication. Many reference books record the powder microscopic characters of herbal drugs. For example, Chinese Materia Medica, European Pharmacopoeia, British Pharmacopoeia, United States Pharmacopeia, Japanese Pharmacopoeia, Ayurvedic Pharmacopoeia of India and Vietnamese Pharmacopoeia.

Botanical microscopy is a unique, valuable, rapid and cost-effective assessment tool. It continues to play an important role in the authentication and assessment of medicinal plants. American Herbal Pharmacopoeia (AHP) gives special emphasis on the powder microscopy of the herbal plants.

References

1. D.B.A. Narayana, Pharmacognosy and technology, Pharmacognosy Magzine 6 (23) (2010) 145–146.
2. World Health Organisation, Quality Control Methods for Medicinal Plants, 1998. Geneva.
3. World Health Organisation, Traditional Medicine Strategy, 2002–2005. Geneva.
4. European Medicine Evaluation Agency, Guidelines on Quality of Herbal Medicinal Products, 2006. EMEA/CVMP/814/00. London.
5. World Health Organisation, Guidelines on Good Agricultural and Collection Practices(GACP) for Medicinal Plants, 2003. Geneva.
6. P.K. Warrier, V.P.K. Nambiar, C. Ramankutty (Eds.), Indian Medicinal Plants, vols. 1–5, Orient Black Swan/Universities press, 1994.
7. V.V. Sivarajan, I. Balchandran, Ayurvedic Drugs and Their Plant Sources, Oxford & IBH Publishing CO. pvt Ltd, New Delhi, 1994.
8. B.L. Vaidya, Some Controversial Drugs in Indian Medicine, Chaukhambha Orientalia, Varanasi, 1982.
9. K.M. Ganeshaiah, R. Ganeshan, R.U. Shankar, C. Meera, A taxonomic hurdle by taxonomists, AMRUTH (1998) 3–8.
10. M.D. Rajendra, D.B. Ananthanarayana, Controversial nomenclature of ayurvedic drugs. Challenges for scientists, Pharmacognosy Reviews 3 (5) (2009) 1–7.

11. D.B.A. Narayana, Pharmacognosy and technology, Pharmacognosy Magazine 6 (23) (2010) 145–146.
12. The Editorial Committee, Materia Medica of China, Shanghai Science and Technology Publishing House, Shanghai, 1999.
13. European Pharmacopoeia Commission, European Pharmacopoeia, sixth ed., European Directorate for the Quality of Medicines and Healthcare of Council of Europe, Strasbourg, 2007.
14. British Pharmacopoeia Commission, British Pharmacopoeia, The Stationery Office on Behalf of the Medicines and Healthcare Products Regulatory Agency, London, 2008.
15. The United States Pharmacopeial Convention, The United States Pharmacopeia, in: 30th Revision/National Formulary, twenty-fifth ed., The United States Pharmacopeial Convention, Rochville, 2005.
16. Japanese Pharmacopoeia Committee, The Japanese Pharmacopoeia, fifteenth ed., Society of Japanese Pharmacopoeia, Tokyo, 2006.
17. The Committee of the Ayurvedic Pharmacopoeia of India. The Ayurvedic Pharmacopoeia of India. New Delhi: The Controller of Publications Civil Lines.
18. Vietnamese Pharmacopoeia Commission, Ministry of Health. Vietnamese Pharmacopoeia, third ed., Ministry of Health – S.R. Vietnam, Hanoi, Vietnam, 2005.
19. American Herbal Pharmacopoeia, Roy Upton Herbalist, 1999.

Chapter 2

Grinding of plant materials

2.1 Need for grinding of herbs

For microscopic and other such testing, herbs need to be in either cut or powdered form. In the laboratory, depending on its nature, small quantities of samples of the herb can be ground using a domestic dry grinder fitted with blades; or in case the herb is hard, it may be ground by pounding using a mortar and pestle. Porcelain, iron, and wooden mortar and pestles are common. Mortars are also made of agate or granite. However, in any industrial setup or even for traditional healers, vaidyas would need to grind large quantities of herbs to prepare herbal products or recipes to dispense to a large number of patients.

The herbal industry has grown over the past decades; it supplies not only raw unprocessed herbs but also processed herbs for use in the industry. Such processing is carried out on the scale of a few kilograms to hundreds of kilograms and tons. The grinding of herbs obviously provides a large surface area, which makes it convenient to prepare or manufacture dosage forms for oral or external application, for extraction with solvents from water or mixtures of water and ethanol and other organic solvents. The latter is referred to as extracts and may be a soft extract, liquids, or dried powders obtained by technologies such as freeze drying, forced air drying, spray drying, or adsorbing the liquid extract onto an inert substance and drying the adsorbed material. Enhancement of the surface area is an important step to enhancing the extractive values of the herb. This step also makes it useful for formulating and processing a dosage form so that a defined unit quantity of the herb can be provided for easy administration. For example, traditional knowledge Ayurveda has a dosage form named *churna*, which means powders for oral administration [1]. Other dosage forms such as tablets, pills, hard or soft capsules, pastes, gels, and granules also need herbs in ground form for blending with other additives and excipients during manufacturing. Many herbs are supplied in bulk and need to be chopped or cut into a size suitable for feeding into a grinder. Some of the herbal industry has opted to procure already processed ground herbs rather than buying unprocessed herbs and grinding them in house. The primary reason stated by stakeholders is that grinding operations are messy and dusty and risk contamination with dust.

Powdered Crude Drug Microscopy of Leaves and Barks. https://doi.org/10.1016/B978-0-12-818092-1.00002-6

Moreover, they need large spaces and are energy-intensive operations, which adds to the cost. Procuring ready-ground herbs allows an herbal industry to operate with greater cleanliness and the ability to comply with good manufacturing practices (GMPs).

Hence, at the commercial level, grinding operations have a number of considerations that have a bearing on the design and layout of the grinding parts, the taste of the herbs, and their physicochemical nature.

2.2 Classification of herbs with reference to good manufacturing practices

Herbs and their grinding section need to be completely isolated from other sections of the production unit, because grinding operations often generate large amounts of dust that have potential to contaminate other areas of the building. Herbs can be classified according to their taste, colour, potential for irritation, potential for allergenicity and physical nature, such as soft, hard, fibrous and sticky. Natural chemical constituents such as stickiness caused by their fat or oily contents may make it difficult to grind, because volatile oil contents may evaporate during grinding owing to friction and heat that is generated, and so on. GMP considerations need to be applied when handling herbs to avoid cross-contamination, contamination with other production units when in the same premises, and contamination with the environment and surrounding areas as a result of fine dust allowed to escape from the grinding rooms, and to reduce and minimise potential health hazard to the workers.

With the exception of general references to keeping clearly demarcated areas and preventing cross-contamination, the author has not seen more detailed guidelines and descriptions related to grinding operations in any GMP document including the World Health Organization GMP document for herbal materials. Because the chapter deals with the aspects of grinding, detailed discussions on GMP related to herbs are not covered here.

2.3 Classification of herbs with reference to taste

It is known that herbs have different tastes. The existence of six different tastes is described in Ayurveda as well as in current knowledge [2]: sweet, sour, salty, pungent, astringent, and bitter. Table 2.1 provides a non-exhaustive short list of herbs according to their taste.

Using the same grinder for herbs of different taste has the high potential to contaminate and spoil the nature and constituents of ground herbs. Some of the taste attributes may contaminate the first few kilograms of the ground herb that has a different taste whereas the latter midstream and end-stream powders may not show any contamination. Data on the ability to remove the taste of the herb or the lowest quantity that can still taste like the previously ground herb (bitter or pungent, for example) are not unavailable and need to be generated. In-

TABLE 2.1 Herbs with different tastes.

S.no.	Name	English name	Botanical source	Taste
1	Cinchona	Cinchona bark, Peruvian bark, Jesuit's bark	Cinchona bark is the dried stem and root bark of cultivated trees of *Cinchona calisaya* Wedd., *Cinchona ledgeriana* Moens, *Cinchona succirubra* Pav. Ex. Klotzsch, and *Cinchona officinalis* Linn. or of hybrids of either of the last two species with either of the first two, **Fam**: Rubiaceae	Intensely bitter and astringent
2	Cinnamon	Ceylon cinnamon, Cinnamon	Cinnamon is the dried inner bark of coppiced trees of *Cinnamomum zeylanicum* Blume., **Fam**: Lauraceae	Sweet, agreeable, and warm
3	Cannabis	Indian hemp	Cannabis consists of the dried flowering tops of the cultivated pistillate plants *of Cannabis sativa* Linn., **Fam**: Cannabaceae	Bitter
4	Clove	Cloves	Clove is the dried flower buds of *Syzygium aromaticum* (Linn.) Merr & L.M. Perry. (Syn: *Eugenia caryophyllus* Sprengel), **Fam:** Myrtaceae	Pungent, warm, and aromatic
5	Black pepper	Black pepper	Pepper is the dried unripe fruit of *Piper nigrum* Linn., **Fam:** Piperaceae	Pungent
6	Lemon peel	Lemon peel	Lemon peel is the fresh or dried outer part of the pericarp of	Bitter

Continued

TABLE 2.1 Herbs with different tastes.—cont'd

S.no.	Name	English name	Botanical source	Taste
			the ripe or nearly ripe fruit of *Citrus limon* (Linn.) Burm. f., **Fam**: Rutaceae	
7	Fennel	Sweet fennel	Fennel consists of the dried ripe fruits of *Foeniculum vulgare* Mill, **Fam**: Apiaceae	Sweet and agreeable
8	Eucalyptus	Eucalyptus	Eucalyptus consists of mature whole or cut leaves of *Eucalyptus globulus* Labill, **Fam**: Myrtaceae	Pungent, camphoraceous
9	Ginger	Dry ginger	Ginger is the dried rhizome of *Zingiber officinalis* Roscoe., **Fam**: Zingiberaceae	Pungent
10	Liquorice	Liquorice, glycyrrhiza	Liquorice consists of the dried peeled or unpeeled root and stolon of *Glycyrrhiza glabra* Linn., **Fam**: Fabaceae	Sweet
11	Turmeric	Turmeric	Turmeric consists of fresh or dried rhizomes of *Curcuma longa* Linn., **Fam:** Zingiberaceae	Aromatic, slightly bitter

Fam, family.

house grinding sections of a unit that handle herbs of different taste need to develop standard operating procedures and comply with them for changeover.

Suppliers of herbs in ground form normally adopt dedicated grinders for herbs of a particular taste. Adopting a well-developed cleaning and changeover process and a step to validate the removal of the previously ground herb can also meet GMP requirements if large-scale grinding is undertaken. Such a step is often adopted to optimise investment and operational costs. Validation in such cases often presents serious challenges because most grinders cannot be washed with a vehicle such as water or soapy water. Moreover, in most countries, test tasting is not allowed or part of protocol, to protect humans from such tests. One option is to sample material that adheres to the grinder and examine it using high-performance thin-layer chromatography testing, which detect materials at parts per million levels of compounds.

2.4 Classification of herbs with reference to nature and contents

Physicochemical features of herbs have a direct bearing on the type of grinder and grinding mechanism. The nature of herbs can be different: soft, leafy, woody, or fibrous, with scales of hardness up to extremely hard and dry (such as some heartwood, bark, and nuts). Other features should be considered to select the correct type of grinder and grinding conditions: for instance, for herbs containing volatile oils, fixed oils, gums, resins, or fatty substances. Fig. 2.1 shows a broad classification of softness to hardness.

Herbs are classified according to whether they are highly starchy (Amrita/Giloy: *Tinospora cordifolia*, stem and root), contain volatile oils (cumin seeds: *Cuminum cyminum*), contain fixed oils and fatty substances (almond and other edible nuts, castor roots and fruits: *Ricinus communis*), have a high fibrous content (dried ginger: *Zingiber officinale*), or are pungent (black pepper: *Piper nigrum*).

Most of the grinding requirements are applied to either air-dried or sun-dried herbs needed in powdered form for processing for the needs described in the section above in this chapter. Here, "dry" refers to herbs that have an inherent moisture level normally not exceeding 7–8% w/w when tested for loss upon drying at 105° ± 2 °C. Some suppliers may supply herbs with up to 15% w/w moisture; these would need prior drying before they are ground, for better and smoother operations.

A small number of recipes or formulations use wet ground paste to obtain juices for incorporation into the formulations or for direct administration.

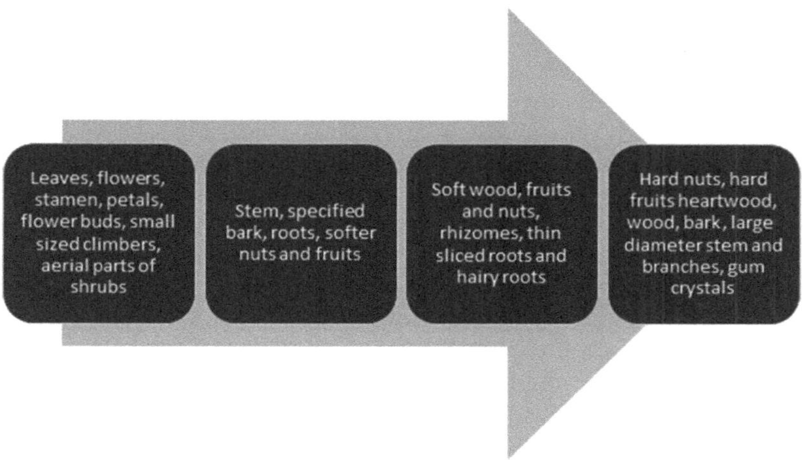

FIG. 2.1 Proposed grid classifying plant parts basis hardness.

2.5 Considerations for design and layout of grinding sections

a. The design of the grinding area needs careful considerations regarding the flooring, walls, roof, and windows or exhaust supplies.

b. Adequate partitioning between two different grinders should be provided.

c. The surfaces of such partitioning and walls must be made suitable for easy dusting and should be dust resistant when possible; they should be easily cleaned by washing and mopping.

d. The surface may be made suitable with washable oily paint surfaces or epoxy coating, or with ceramic tiles in which the joints are well-sealed.

e. The flooring can be made of natural granite, stone, or tiles. Although these materials provide a clean facility they have the disadvantage of being slippery, especially when wet, and are potentially hazardous to workers. Current technology involves epoxy coating the floor, which is an expensive investment compared with other materials.

f. The walls and flooring should be designed with no crevices or joints where dust can accumulate and lead over time to potential fungal and microbial growth.

g. Wherever windows are provided, they need to be airtight, fixed, and double-walled with no crevices and aligned to the surfaces of the wall.

h. The height of the grinding area normally needs to be higher than that of the normal production area to provide adequate space for hanging dust collector bags and filter bags. Taller rooms are preferred to reduce dust levels on the floor. Avoid false ceiling in the roofs because they often provide space for dust to settle and make it difficult to clean. The surface of the roof should be similar to that of the walls and floor described earlier.

i. Considerations are needed for the design of air handling units to provide for air circulation and collect pumped-in air and for a number of air changes. Except when grinding herbs that require a temperature below 25°C, adequate air circulation need to be built in to provide for a convenient working environment for workers. Return air for air handling units should be taken from the lower level (i.e., through return air risers, because dust always tend to settle downwards).

j. Consideration for dust collectors and dust removers needs highest attention from all angles. Simple large exhaust fans installed just below ceiling levels normally are adopted to provide the minimum level of control. Specifically, dust collectors and extractors need to be designed and installed on a case-to-case basis. Cartridge filter bags of a maximum 5-μm pore size with pulse jet cleaning can be a better option in place of simple fabric bags. This will provide better efficiency because adhering dust will be removed periodically from the filter bags and collected in the container below them.

k. In specific cases in which the herb is coloured, special considerations are required. Over a period, despite cleaning, the walls are known to take on the colour of the material and it will be difficult to remove the colour completely (for example, turmeric [*Curcuma longa*] rhizome powdering and red chilies [*Capsicum annuum*]). In such cases, it may be desirable to create a dedicated separate grinding area for such herbs.
l. Care should be taken to design filters for air that is exhausted from grinding areas causing contamination of the external atmosphere of the unit. Dust exhaust has been known to settle slowly and collect onto other areas of the factory or nearby buildings.
m. The loading bay and dispatch bay need to be planned properly with self-closable doors and air curtains when possible. Alternatively, strip polyvinyl chloride curtains with air curtains may be employed.
n. Although the effect of UV light is known to lessen owing to dust, it may be appropriate to install a number of UV tubes at the ceiling in addition to standard illuminating lights, documenting their usage hours. Switch these UV lights on when grinding operations are not ongoing or overnight when there is no work shift so that the microbial load within the area can be reduced and prevent workers from exposure to UV. Direct exposure of UV lights to human eyes must be avoided.
o. There also needs to be an adequate plan to have low humidity in the grinding area; this can be achieved by installing dehumidifiers. A relative humidity of 50% ± 5% relative humidity would be a good target to achieve.
p. Suitable facilities would need to be created to sprinkle fine droplets of water mixed with fumigants such as 0.5% benzalkonium chloride or other approved fumigants. Such fumigants should sprayed starting from ceiling height, allowing droplets to fall slowly to the floor. After 15–30 min of exposure, the floor needs to be mopped. Fumigation with hydrogen peroxide-based fumigants such as Silvicide, germicide can also be a good option. Disinfection of equipment with the help of ozone gas purging can also be explored.
q. Workers need to be provided with changing rooms, washrooms, safety eye wash units, safety sprinklers apart from specifically designed clothing, and head and facemasks to protect them. In specific instance, appropriate nasal masks with filters should be provided and worn to filter dust particles and protect workers' lungs.
r. Other general aspects of GMP need to be applied and adhered to.

2.6 Unit operations involved in grinding

2.6.1 Cleaning of herbs

The author, who has had much experience in traditional medicine as well as in the industry and processing herbal products, concludes that there are a number of reasons why industries adopt different methods that result in unclean herbs. Most often, no cleaning is done. When it is performed, trained workers sit on the floor with the herb spread on sheets on the floor. The workers visually check portions of the supply, remove foreign matter, and collect cleaned herbs and foreign matter separately in bags. Most often this is done manually. More recently, big hoppers or loading areas are connected to long conveyer belts on which the supply moves slowly. The belt design changes from plain to perforated to mesh so that during movement, small adhering particles and small-sized foreign matter fall below the belt. Workers stand on either side of the belt and pick up undesirable and foreign matter and collect them separately. At the end of the belt, the cleaned herbs are collected into containers for the next step. Industry sets descriptions of what constitutes foreign to be removed for each herb.

The food industry, which also uses plant materials, has seen developments in automated machines and technology for cleaning herb-based raw materials: for example, an automated line involving conveyer belts, roller brushes, and forced air blowers are used to clean dry garlic bulbs. These remove foreign matter as well as dried unusable scales sticking to garlic bulbs.

Similarly, vibrating equipment with sieves of different mesh sizes are commonly used to clean grains and pulses to remove foreign matter as well as stones and soil. The herbal industry needs to adopt such equipment for herbs to increase efficiency and productivity and comply with GMP by using cleaned herbs.

2.6.2 Washing and drying

The author with long experience in the traditional medicine as well as herbal products processing and industry has been exposed to citing number of reasons that drive industries to adopt different methods not to wash and dry the herbs prior to grinding or further processing. Often, herbs are not washed before they are ground or further processed. Cited reasons cited an inadequate water supply, the highly energy-intensive steps needed to dry the herbs, unavailability of continuous drying technology, the very slow process of drying if tray driers are used (leading to batchwise operations), the need for large human resources and associated costs, and inadequate knowledge about the effects of water used on the surface of the herb and on the quality of the herb. A section of the traditional medicine industry in India is known to load herbs into large tote boxes (containers with all five sides fitted with perforated sheets). These tote boxes normally have castor wheels and are pulled to a washroom with a rooftop tap with sprinklers. Water is sprinkled onto the contents of the tote boxes for a few minutes and excess water is allowed to drain away. Then these boxes are pulled to a large outer room fitted with hot air circulation to dry or are exposed to an open area for sun drying. Implementing such washing and drying steps in advance to obtain washed and dried herbs for grinding or other processing steps should form part of production planning.

Big hoppers or loading areas have been designed that are connected to long conveyer belts on which the herb supply moves slowly. On the belt, a number of water sprinklers and jets are fitted through which water is supplied at a predetermined speed and pressure. The sprinkled water washes the herbs efficiently and the lower tray below the belt collects the washed water with the dirt. The belt design changes from perforated to sieves with a mesh so that during movement, small particles and foreign matter fall below the belt as washing takes place. Some of the washing lines are also fitted with roller brushes through which the herbs pass, causing abrasion and improving the efficacy of cleaning. At the end of the belt, the washed and cleaned herbs are collected in containers for the next step or for the next section of the conveyer. This section is often a tunnel-like structure fitted with a heating system with or without air blowers to remove surface moisture rapidly and dry the material.

The food industry has seen developments in automated machines and technology for washing herb-based raw materials. For example, an automated line involving conveyer belts, roller brushes, jet sprinklers, and forced air blowers is used to wash garlic bulbs, tomatoes, mangos and such other fruits, spinach, tea leaves, ginger, and green chilies. It removes foreign matter and provides excellent washing.

The herbal industry needs to adopt such equipment for herbs to increase efficiency and productivity and to comply with GMP by using cleaned and washed herbs.

At a fundamental level, the need to wash herbs is a point of discussion. The pharmaceutical and drug industries desire and perhaps insist on highly efficient washing of the herbs to remove adherent matter and all associated microbial flora. Traditional industry experts and vaidyas opine that it is not the herb alone but also the associated microbial flora that have an important role in the safety and efficacy of the herb. Hence, for many herbs traditional vaidyas do not want to provide a high level of washing. High surface fungal contamination leading to potential aflatoxin formation that can further contaminate the batch is an issue that needs to be addressed. These aspects are interesting subjects for the study and development of appropriate technologies. Experiments involving ozonised water, water with sanitisers such as silver solutions, and warm water for washing can lead to improved efficacy of surface washing.

2.6.3 Wet grinding and juicing

Wet grinding is adopted to obtain a paste or gel mass from fresh herb, a paste mass using a dried herb, or juice. Traditional processing uses the term *swaras*, which refers to the juice of herb in question required for mixing with or trituration of other herbs. Such *swaras* is normally obtained from fresh herbs. Ayurveda also permits juice to be generated from dried herbs by soaking them with a prescribed quantity of water and mildly grinding them followed by straining.

Edge runner mills, end runner mills, motor-driven large mortars and pestles (in which the mortars and pestles rotate in opposite directions), and reciprocating flat-bedded mortars and pestles are commonly adopted for such wet grinding. The quality of herbs, and especially their particle size, for wet grinding needs be determined for each herb. The quality of the menstruum (water or any other vehicle) also need to be specified. The debate cited earlier related to the microbial load normally present on the herb taken for grinding applies in this step as well. One side feels no alteration of the microbial flora should be attempted whereas another school of thought thinks that all adhering microbial flora should be completely eliminated.

2.6.4 Dry grinding

Dry grinding is the process adopted for most herbs owing to the needs covered earlier. The classification of herbs based on various attributes was covered in other sections of this chapter.

For soft herbs, a number of options are available for grinding. These include the use of a multimill with the blade forward and in some cases with the hammer forward, crushing mills, roller stone mills, and other grinders that provide a circular grinding mechanism. Most such grinders have a sieve fitted so that the ground material passes through it; material that has not attained the particle size continues to be ground until it passes through the sieve.

For soft herbs that are highly fibrous, these types of grinders do not work. For such herbs, specially designed jaw crushers are used. Such crushers provide a strain and pulling mechanism so that the fibrous material is broken into smaller lengths and then powdered. For example, most rhizomes and roots such as ginger and licorice (*Glycyrrhiza glabra*) need such grinders.

For mild to moderately hard herbs (such as pepper, pippali [*Piper longum*] and nutmeg [*Myristica fragrans*]), the stems of some plants, pieces of heartwood, and dried fruits such as myrobalans, multimills with the blade forward and in some cases with the hammer forward or high-speed hammer mills are often appropriate equipment.

For hard materials such as bark, hard heartwood, roots, root bark, hard nuts, and fruit, specially designed hammer mills with strong hammers and a relevant speed of rotation are available.

Because of friction, most of this equipment generates heat; a transient temperature rise of 5–10°C is normally seen in the output ground powder. The powder needs to be exposed for a few minutes to attain room temperature; otherwise, it tends to agglomerate owing to what is commonly called sweating. Such an increase in temperature causes frequent blocking or stoppage of grinders, especially if the herb contains oily or fatty constituents. In the latter case, the grinders often jam; in such cases, appropriate technologies need to be developed. These grinders need to be fitted with sieves to obtain a coarse or fine powder, according to the need. The finer the mesh size of the sieve, the greater would be the difficulty of grinding and the slowing the grinding steps.

The generation of heat and its impact on the aroma and flavor of the ground herbs are a matter of discussion and of great interest to consumers. According to tradition, people believe that pounding with mortars and pestles (iron or wooden) provides much better quality of the ground herbs. In fact, this is a point of debate for many types of grinding of both dry and wet materials. Some companies that supply pre-ground herbs, especially spices and spice mixes for use in food preparations and for seasoning, claim that their powders are "hand pounded." It is possible to perform such pounding on a commercial scale because many pestles may be connected to a single rod that sequentially move up and down a shaft and pound the material in individual mortars kept below. Such machines are custom-made and used to process resinous materials such as guggul (*Commiphora mukul* and other species) to be ground with other extracts.

The feed material of the herbs needs to be of a size suitable for feed into hoppers. In small quantities, they are hand chopped; otherwise, machines used to saw wooden trunks are adopted. Grass choppers are ideal for grass-like materials such as stems, leaves, and aerial parts.

2.6.5 Roasting and grinding

In specific cases, it has been found advantageous to take the raw herb through mild to moderate roasting steps before grinding. Roasting removes most of the remaining water and alters the presence of volatile oil from inside to outside the cells, providing higher brittleness that support easier grinding. Roasting can also alter the chemical composition; hence, care should be taken before applying roasting as a pre-grinding step. Rotating drum roasters, open pan roasters, vacuum roasters, and simple roaster conveyer belts are available in which the material is exposed to heat from the top or bottom of the conveyer belt.

Roasting is often used for a number of herbs such as chicory roots and stems, fenugreek seeds, and cumin seeds.

2.6.6 Sifting and separation based on particle size

All ground herbs should be passed through appropriate sieves using sifters. This step is important for maintaining the particle size range of the powdered herb and remove fines and inadequately ground materials. Sifting often retains some unground and foreign matter as well as metal pieces if are present or generated during grinding, and eliminates them.

Vibratory sifters are often used; they are available in various sizes, diameters and speeds. Large vibratory sifters with multiple sieves fitted are also available that separate the powdered herb into different particle size ranges. Although such sifters are uncommon in the herbal industry, they are used globally to sift ground ispaghula (*Plantago ovata* seed and husk). Ground ispaghula is portioned into different particle size ranges to provide for coarse ground ispaghula and finely ground and grit-free (providing better mouth feel) ispaghula, which goes into formulating powdered dosage forms or granulated dosage forms used to treat constipation and irritable bowel syndrome. Such a system needs to be evaluated for ground herbs used to formulating *churnas* or powdered dosage forms directly administered orally. Such a ground herb would also form an excellent starting raw material owing to its large surface area for converting into pastes or gels for external application or for unit dose (specified spoon quantities) for oral administration.

2.7 Cryogrinding

Cryogrinding refers to grinders in which the grinding chambers have an outer jacket that creates a low-temperature condition in the grinding chamber. The temperature inside the grinder can be fixed at any desired low temperature below 15°C to less than 0°C. The lower the temperature, the lesser would be the effect on the herb owing to heat generated during grinding. Cryogrinding is recommended for herbs that have inherent volatile oils, fixed oils, fats, and oleoresins. It is known to cause freezing and potential ice formation, which making the herb hard at that temperature and leads to potential sticking and jamming of the grinders. Careful experimentation to set the optimum cooling of the grinding chamber is an important step to grind such herbs without loss of volatile or fixed oil contents. There are only few examples of herbs ground using cryogrinding by the herbal and traditional medicinal industry. Grinders fitted with water-cooled jackets are available and the possibility of using such equipment may be explored.

2.8 Jet mill grinding

Jet mills are commonly used to micronise minerals and synthetic drugs. They operate by throwing material already in a coarsely ground form at high speed onto a hard surface with great force. The impinging effect breaks down the particles into micronised sizes (to 100 μm or less). They can also be used with appropriate adoption to obtain herbs ground into micronised sizes for better absorption, to improve the surface area and hence provide multiple advantages. Herbal powders that need to be orally administered directly or powders that need to be made into a gel or paste for external application on skin, wounds, head, hair, and joints can be jet milled for improved benefits.

2.9 Dust control, safety, health and environment

Many aspects of dust control have been discussed in earlier sections of this chapter. It is important to do so effectively not only from an environment point of view but for the highly significant viewpoint of the safety and health of workers. In some cases, there is prior knowledge of a potential allergy to workers, as in the case of various nuts such as almonds, cashews, walnuts, and groundnuts (peanuts). Data on the irritancy potential are common. However, allergenicity can happen to anyone from any herb that is processed for grinding. The continuous recording of data of such observations should be made at the individual level and the common level (when an event happens to all who are exposed to that herb while grinding). Such records should be updated periodically and applied rigorously. Medical attention ready to be applied for emergencies is necessary.

Many herbs that need to be ground may have thorns or sharp trichomes, and sharp cut edges that can cause injury or reactions to the skin. Pollen grains from flowers are another source of such adverse reactions: for example, datura (*Argemone mexicana*) and rose stems. A classic case known in Ayurveda is the use of the fruits of *bhallataka* (*Semecarpus anacardium*). The fruit of this herb is irritating to handle without pretreatment. It can cause a severe reaction and in some cases if untreated may quickly be fatal. The fixed oil present in the pericarp of this fruit is the source of the problem. According to the traditional process, the fruit is to be boiled as such (without grinding) in milk and then dried. The dried fruit can be easily handled with no reaction. Another process is to keep the unground fruits in a paste of mud or brick powder with some water overnight, washed and dried. Such treated fruit of *bhallataka* can easily be handled with no reaction.

2.10 Packing of ground herbs and storage

Ground herbs are normally packed based on the need and final use. If they are needed for use by vaidyas in clinics and pharmacies, they may be packed in laminated bags, poly bags, self-sealable plastic bags, aluminum foil bags with multiple laminations, and or bottles (glass or high-density polyethylene [HDPE]) with lids. These need to be sealed effectively and provide protection from moisture and air. Nitrogen flushing of these packs can be considered when necessary. These packs can be from a few grams to a few kilograms.

Bulk packs are normally in large double poly bags or woven sacks whose mouths are twisted and tightly tied with threads or rubber bands. The bags are then kept in outer drums (made of boards or HDPE) with tightly closed lids and rims to seal them.

Avoid using stapler pins or any such devices in packing. These packs are normally anywhere from 10 to 300 kg, depending on the bulk density of the ground powders. All such packs need to be appropriately labelled with the requirements of national regulations.

If the hygroscopicity of the ground herbs is known, consider adding and blending well a permitted anticaking agent such as calcium carbonate or colloidal silicon dioxide before packing, as permitted in the country. Labels should then declare such an addition.

2.11 Microbial quality considerations

It is important during grinding operations to give adequate considerations to maintaining or reducing the microbial quality basis the buyer's specifications. Most pharmacopeia and World Health Organization documents provide quality specifications that prescribe acceptable levels of total plate counts (TPCs), fungal counts, yeast and molds, coliform organisms, and the absence of *Escherichia coli* and other pathogenic organisms. Different limits are specified for those to be used as powders and those for use as teas or decoctions after boiling. Some buyers demand the TPC to be below 1000 colony-forming units, which is difficult to achieve. Technologies to treat packs of ground herbs for exposure to low-dose gamma radiation or ethylene oxide (ETO) fumigation are available. Such gamma radiation treated packs and powders need to declare the same on the label with specific pictorial symbols, which raises eyebrows from the users causing reluctance to use. ETO-fumigated herbs, on the other hand, need to be tested for residual ETO levels and permitted levels are very low. ETO residues may remain in pockets in the pack although testing may meet the requirements in the sample that is drawn, which concerns many users carry because ETO residues can cause carcinoma. Long-term exposure to low-temperature heating is normally recommended; however, its efficiency is not definitely proven and it can vary in the requirements from herb to herb. This process is also highly energy intensive and expensive.

2.12 Disinfection of ground herbs

Ground herbs are more prone to bacterial growth because of the larger surface area. The possibility of disinfection with ozone gas-based disinfecting agents may be explored before packing. This is an area for the development of newer technologies that are cost-effective and implementable.

2.13 Cleaning and maintenance of grinders

Preventive and periodic maintenance of grinders and their parts is mandatory. Hammers, blades, cutting edges, stone, agate, iron and bronze grinding plates, motors, mortars and other parts need to be checked, repaired and replaced. Fitted sieves need checking sometimes after each use for breaks that cause metal filings, pieces, and metal dust contamination with the ground materials.

Most grinders cannot possibly be cleaned using any vehicle such as water, for obvious reasons, except where parts that come into contact can be detached from the machine and can be washed with both detergent water and clean water and dried before fitting into the machine. Wiping parts with a cloth wet with isopropyl alcohol is another option. If the grinders are in dedicated use, they may not need cleaning except during maintenance operations. It is best to pass a small quantity of inert materials such as starch powder, microcrystalline cellulose powder, calcium carbonate or calcium phosphate di- or tribasic through the grinders repeatedly to clean them. Exhausted herbal powders such as the powders of herbs that have been processed through extraction (most soluble ingredients are already extracted) and dried are another option. Sawdust may be also considered. Studies need to be undertaken before selecting and adopting such cleaning materials.

2.14 Ensuring metal-free ground herbs

Adopting most of the conditions and steps described in this chapter can reduce the presence of metal dust and pieces in the ground produce. Such contaminants are not acceptable; hence, metal removers should be used before packing the ground herbs. Equipment can magnetically catch such contaminants and remove them. The challenge is in metals of a nonmagnetic nature. It is best to have the final step in which packs pass through an efficient metal detector before approving the consignment and to remove packs that show contaminants for reprocessing.

2.15 Future developments

There is a need to develop technologies for sterilization and bio-load reductions, grinding of herbs with volatile oil and fat content, equipment that can reduce high energy consumption and technologies that do not require excess water requirement during washing. High-efficiency dust collectors and removers constitute another area. Detailed broad general monographs on the grinding of herbs do not appear in pharmacopoeias. Official guidelines and a general consensus on this area are also not seen in regulations. Regulators need to consider beginning work in this area.

References

[1] Ayurvedic Formulary of India, Ministry of Health and Family Welfare, second ed., Government of India, 2003. Part I, 103–119.

[2] R.amachandra Reddy, et al., Bheshaj Kalpana Vignanam Chaukhambha Sanskrit Bhawan, in: Varanasi, India, fourth ed, 2015, pp. 41–42.

Chapter 3

Microscopic identifying features of barks and leaves

A bark powder is identified microscopically by various features such as stone cells, calcium oxalate crystals, starch grains, medullary rays, fibres, sclereids, cork, isolated oil cells, tubular lactiferous canals, phloem parenchyma, masses, rhytidoma, parenchyma and secretory canals. Bark powder cannot be merely authenticated by observing the presence or absence of these features as these are common in all barks. The variation lies in the size and shape of the stone cells, sclereids and fibres, their occurrence, type of wall and lumen, type of calcium oxalate crystals, shape and type of hilum of starch grains.

The features selected for identification of bark powder included 9 general characters namely calcium oxalate crystals, starch grains, medullary rays, cork, tubular lactiferous canals, phloem parenchyma, masses, rhytidoma, and secretory canals and 3 specific characters namely stone cells, sclereids and fibres. These specific characters are further differentiated into 12 features and 127 subfeatures. The powder microscopy of bark was studied based on specific identifying features and their subfeatures as shown in Table 3.1.

The features selected for identification of leaf powder included 11 general characters namely epidermis, calcium oxalate crystals, fibers, stone cells and sclereids, cell contents, cystoliths, lamina, starch grains, tracheids, lactiferous canals, and xylem vessels and 2 specific characters namely stomata and trichomes, which are differentiated into 56 features. The powder microscopy of leaves was studied based on parameters as shown in Table 3.2.

The features selected below will be exemplified by the respective crude drug microscopy in subsequent chapters.

Powdered Crude Drug Microscopy of Leaves and Barks. https://doi.org/10.1016/B978-0-12-818092-1.00003-8

TABLE 3.1 Parameters selected for the study of powdered bark samples.

Stone cells				Sclereids			
Occurrence	**Walls**	**Shape**	**Lumen**	**Occurrence**	**Walls**	**Shape**	**Lumen**
N/A	N/A	N/A	N/A	N/A	N/A	N/A	N/A
Isolated	Thick walled	Oval	Wide	Isolated	Thick walled	Oval	Wide
Group	Lignified	Circular	Narrow	Group	Lignified	Circular	Narrow
Abundant	Striations	Isodiametric	Pitted	Abundant	Striations	Isodiametric	Pitted
Few	Radiating pits	Rectangular	Slit like	Few	Radiating pits	Rectangular	
	Projections	Rounded			Projections	Rounded	
	Crenate margin	Polyhedral			Peg like extension,	Polyhedral	
	Beaded	Spherical			Beaded	Spherical	
	Pitted	Pentagonal			Branched	Pentagonal	
	Transversely pitted	Triangular			Three sided thickening,	Triangular	
	Branching pits	Squarish			Distinct pits	Cylindrical	
	Distinctly pitted	Irregular			Simple pits	Elongated	
	Devoid of pits	Oblong			Pitted walls	Fibre like	
	Thin walled	Angular			Thin walled	Very long	
						Blunt end	
						Pointed ends	
						Oblong	
						Cone shaped	

Fibres				Starch grains			
Occurrence	**Walls**	**Shape**	**Lumen**	**Occurrence**	**Shape**	**Type of hilum**	**Number of concentric striations**
N/A	N/A	N/A	N/A	Abundant	N/A	N/A	N/A
Isolated	Thick-walled	Oval	Wide	Few	Simple	Narrow	42769
Numerous	Lignified	Circular	Narrow		Round	Cleft	Distinct striations
Group	Striations	Rectangular	Uneven		Oval	Radiate	
Few	Radiating pits	Rounded	Highly pitted wide lumen		Compound	Slit	
	Oblique pits	Polyhedral	Slit like		Spherical	Eccentric	
	Projections	Spherical			Triangular	Centrally located hilum	
	Frequently broken	Pentagonal			Small		
	Occasionally twisted	Triangular					
	Beaded	Squarish					
	Non lignified	Irregular margin					
	Septate	Very long					
	Crenate margin	Pointed apex					
	Smoothly/finely striated	Bent					

Continued

TABLE 3.1 Parameters selected for the study of powdered bark samples.—cont'd

	Fibres			Starch grains			
Occurrence	Walls	Shape	Lumen	Occurrence	Shape	Type of hilum	Number of concentric striations
	Dentate margin						
	Faint transverse septa						
	Thin walled						
	Irregular swellings						
	Crystal fibres						
	Transverse pits						
	Bifurcating ends						
	Distinct pits						
	Slit like						

Calcium oxalate crystals	Medullary rays	Cork	Isolated oil cells	Tubular lactiferous canals	Phloem parenchyma	Masses	Rhytidoma
N/A	N/A	N/A	Ovoid	N/A	N/A	N/A	N/A
Prismatic	Uniseriate	Thick walled	N/A	Groups	Spherical	Granular masses	Irregular
Clusters	Biseriate	Lignified		Oval	Oval	Dark brown cells containing tannins	Rectangular
Rosettes	Multiseriate	Suberized		Spherical	Pitted	Brown gummy	
Acicular	Tangentially-cut	Polygonal		Isolated	Lignified	Circular	
Microsphenoidal	Radially-cut	Isodiametric				Large oval	
Sandy	Longitudinally-cut	Hexagonal					
Needles		Rectangular					
Crystal layer		Pentagonal					
Crystal sheath		Beaded walls					

Continued

TABLE 3.1 Parameters selected for the study of powdered bark samples.—cont'd

Calcium oxalate crystals	Medullary rays	Cork	Isolated oil cells	Tubular lactiferous canals	Phloem parenchyma	Masses	Rhytidoma
Idioblasts		Pitted lumen					
Rod shaped		Oval					
		Elongated					
		Thin walled					
		Multilayered					
		Radially arranged					
		Irregular					

TABLE 3.2 Parameters selected for the study of powdered leaf samples.

Epidermis	Stomata	Trichomes	Calcium oxalate crystals	Fibres	Stone cells and sclereids	Cell contents
N/A	N/A	Abundant	N/A	Isolated	Lignified	Volatile oil
Thick	Anomocytic	Few	Prismatic	Group	Non lignified	Mucilage
Thin	Anisocytic	N/A	Clusters	N/A	Wide lumen	Tannin cell
Beaded	Paracytic	Simple	Rosettes	Thick walled	Narrow lumen	Yellowish pigment
Straight	Diacytic	Branched	Acicular	Thin walled	Rectangular	Fatty oil globule
Wavy	Tetracytic	Covering	Microsphenoidal	Longitudinally cut	Oval	Resin ducts
Polygonal	Cyclocytic	Warty	Sandy	Narrow lumen	Triangular	Brownish matter
Anticlinical walls	Hexacytic	Glandular	Needles	Wide lumen	Fibrous	
Striated cuticle	Actinocytic	Unicellular	Crystal layer	Slit like pits		
Devoid of stomata	Sunken	Thick walled	Crystal sheath			
Slightly sinous walled		Sessile	Idioblasts			
Hexagonal		Multicellular	Rod shaped			

Continued

TABLE 3.2 Parameters selected for the study of powdered leaf samples.—cont'd

Epidermis	Stomata	Trichomes	Calcium oxalate crystals	Fibres	Stone cells and sclereids	Cell contents
Lignified wall		Uniseriate	Diamond shaped idioblasts			
Dome shaped papillae		Biseriate				
Papillose cells		Short stalk unicellular				
Lignified pith cells		Stalk multicellular				
Rectangular		Head unicellular				
		Oval head multicellular				
		Shaggy wooly trichomes				
		Straight				
		Irregularly bent				
		Striated walls				
		Pedestial base				
		Stellate				
		Spear shaped				
		Twisted				
		Narrow neck				

		Oval inflated apical cell				
		Thread like terminals				
		Oval bicellular head				
		Conical				
		Cystolithic				
		Bicellular Cylindrical stalk				
		8-12 celled head				
		Cup shaped head				
		Globular head				
		Pointed apex				
		Short				
		Curved				
		Cicatrix				
		Bent				
		Long				
		Hook shaped				
		Swollen bases				
		Sickle shaped				

Continued

TABLE 3.2 Parameters selected for the study of powdered leaf samples.—cont'd

Cystoliths	Lamina	Starch grains	Tracheids	Lactiferous canals	Xylem vessels
Cigar shaped	Thick walled cells	Few	Spiral	Elongated	Annular
Longitudinally cut	Multilayered epidermis	Abundant	Annular pitted	Tubular	Spiral
Warty	Palisade cells	Simple	Bordered pitted	Yellowish granular latex	Pitted
Cylindrical	Spongy parenchyma	Spherical		Long septate fragments	Reticulate
		Compound			Longitudinally cut
		Oval			Crystal fibres
		Pear shaped			

Chapter 4

Processing and optimized method of preparation of slides

The processing of crude drug and method of slide preparation is of utmost importance to visualize the microscopic characters. It is also important to understand the impact of grinding the crude drug on a lab and industry scale.

Standard operating procedure for preparation of slides of powdered bark and leaf samples.

Two types of slides are to be prepared for the visualization of microscopical features present in a bark. The slide preparation method was optimized to determine the dilution of powdered bark and leaf in water to visualize common as well as distinguishing characters.

Slide I: A 500 mg of moderately fine (44/85) and fine (85) powdered material is to be soaked overnight in 10 mL of water (1:20) for 24 h. Subsequently, the contents are poured in a petriplate and slide is prepared by mounting the contents on a clean and dried slide with a brush. The slide is observed under Motic microscope moticam 3.0 MP, AE 2000.

Slide II: The slide is prepared by potassium chlorate treatment which is used as an oxidizing agent, for disruption of stone cells and sclereids. A 200 mg of powdered material is boiled with 5 mL of 50% nitric acid, followed by a pinch of 100–150 mg of potassium chlorate. The contents are poured in a petriplate after effervescence ceases. The contents are then mounted on a clean and dried glass slide with the help of a brush. The slide is observed under Motic microscope moticam 3.0 MP, AE 2000.

Note: Powdered barks which have high tannin content like *Saraca asoca* and *Syzygium cumini* require oxidative treatment by potassium chlorate for visualization of stone cells and sclereids. *Alstonia scholaris* is a pale buff brown colour powder having low tannin content. Therefore, overnight soaking of powdered bark was enough for visualization of all features including stone cells and sclereids.

The method of preparation of slides was optimized by comparing laboratory pulverized powdered sample with industry grounded samples. The powder microscopy of laboratory pulverized powdered sample and industry grounded

Powdered Crude Drug Microscopy of Leaves and Barks. https://doi.org/10.1016/B978-0-12-818092-1.00004-X

samples was found to be same. However, few observations were found to be as follows:

Calcium oxalate crystals are abundantly present in moderately fine powder (laboratory grinding).

Stone cells and sclereids, the main distinguishing features for any bark are clearer and more prominent in fine powder (industrial grinding) as laboratory grinding is coarse as compared to industrial grinding.

None of the powder characters are destroyed in both moderately fine and fine powder which indicates that grinding technique has no effect on the powder microscopy of the crude drugs.

Reference

[1] D. Singh, V. Aeri, D.B. Ananthanarayana, Development of standard operating protocol for slide preparation of powdered bark samples with varying grinding techniques, Pharmacognosy Journal 10 (2) (2018) 265–271.

Chapter 5

Microscopy/Monograph of powdered bark drugs

The features selected for identification of powdered bark samples includes — stone cells, calcium oxalate crystals, starch grains, medullary rays, fibres, sclereids, cork, isolated oil cells, tubular lactiferous canals, phloem parenchyma, masses, rhytidome, parenchyma, and secretory canals which are further differentiated into various subfeatures. The below mentioned barks are subjected to detailed powder microscopic evaluation using Motic microscope moticam 3.0 MP, AE 2000. The parameters selected for the study of morphology of bark sample are — Shape, colour and texture of outer and inner bark, fracture and odour. The photographs are taken in a grid using Cannon 500D camera with 18-megapixel capacity. The barks selected for morphological and powder microscopic examination are listed below:

5.1 *Aegle marmelos* (L.) Correa
5.2 *Ailanthus excelsa* Roxb.
5.3 *Alstonia scholaris* (L.) R. Br.
5.4 *Anogeissus latifolia* (Roxb. Ex DC.) Wall. ex Guillem. & Perr.
5.5 *Artocarpus heterophyllus* Lam.
5.6 *Bombax ceiba* L.
5.7 *Buchanania lanzan* Spreng.
5.8 *Butea monosperma* (Lam.) Taub.
5.9 *Calotropis procera* (Aiton) Dryand.
5.10 *Carissa carandas* L.
5.11 *Cedrela toona* Roxb. Ex Rottler
5.12 *Crataeva nurvala* Buch. -Ham.
5.13 *Litsea chinensia* Lam.
5.14 *Mangifera indica* L.
5.15 *Moringa oleifera* Lam.
5.16 *Pongamia pinnata* (L.) Pierre
5.17 *Spondias pinnata* (L. f.) Kurz
5.18 *Stereospermum chelonoides* (L. f.) DC.
5.19 *Syzygium cuminii* (L.) Skeels
5.20 *Zizyphus mauritiana* Lam.

Chapter 5.1

Aegle marmelos

Aegle marmelos (L.) Correa Rutaceae

Common vernacular names: Assamese – Bael, Yael; Bengali & Hindi – Bael, Bel; English – Bengal quince; Gujarati – Bill, Bilum; Kannada – Bela, Bilva; Malayalam – Koovalam, Vilvam; Marathi – Baela, Bel; Oriya – Belo; Tamil – Vilvam; Telugu – Bilavamu, Maredu [1] (Fig. 5.1.1).

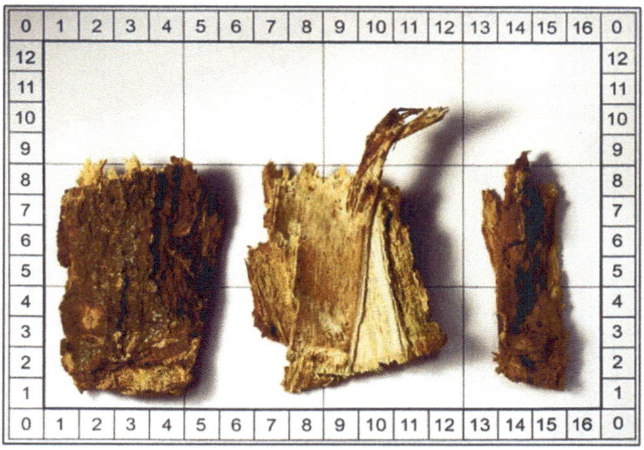

FIG. 5.1.1 *Aegle marmelos* bark.

Macroscopic description: Shape – Flat or curved, **Outer bark** – Greyish brown, longitudinally fissured, **Inner bark** – Buff to pale brown, internally smooth, faintly or longitudinally striated, **Fracture** – Outer short, inner fibrous, **Odour** – Faint and characteristic.

Major chemical constituents: y-Fagarine, marmesin, marrnin, umbelliferone [2].

Traditional uses: Stomach disorder, intermittent fevers, heart disorder [2].

Powder microscopic description: Pale yellowish brown colour powder showing abundant, isolated or groups of characteristically interlocking

Powdered Crude Drug Microscopy of Leaves and Barks. https://doi.org/10.1016/B978-0-12-818092-1.00005-1

sclereids of various shapes, sizes and thickness, few of them have blunt ends, with thick walls, narrow lumen, oblong to oval in shape; Stone cells, in groups or isolated, oval to circular in outline with slit like narrow lumen and with faintly striated walls; Plenty of small sized prismatic crystals; Fragments of tangentially and radially longitudinally cut medullary rays associated with phloem elements and parenchymatous cells; Fragments of very long, thick walled fibres; Polygonal to hexagonal cork cells [3]. The general as well as specific features of *A.marmelos* bark powder are illustrated in Table 5.1.1 and Figs. 5.1.2–5.1.7 respectively.

TABLE 5.1.1 Specific as well as general characters of bark powder of *Aegle marmelos.*

	Specific features			General features
	Stone cells	**Sclereids**	**Fibres**	**Calcium oxalate crystals**
Occurrence	Isolated	Isolated	Numerous	Prismatic
	Group	Group	Group	**Cork**
	Abundant	Abundant		Pentagonal
Walls	Faintly striated	Thick walled	Thick walled	Polygonal
		Radiating pits	Septate	Hexagonal
		Peg like extension		Beaded walls
		Thin walled		Pitted Lumen
Shape	Oval	Rectangular	Very long	**Medullary Rays**
	Circular	Oval		Tangentially cut
	Spherical	Spherical		Radially cut
		Pentagonal		Longitudinally cut
		Blunt end		
Lumen	Narrow	Narrow	Narrow	
	Slit like	Wide		

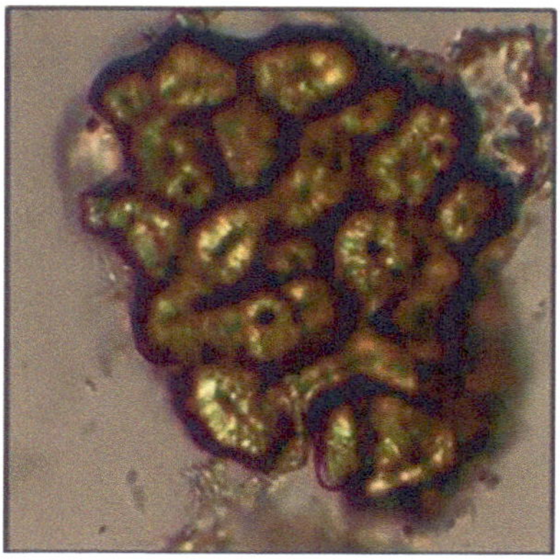

FIG. 5.1.2 Stone cells from *Aegle marmelos* bark powder.

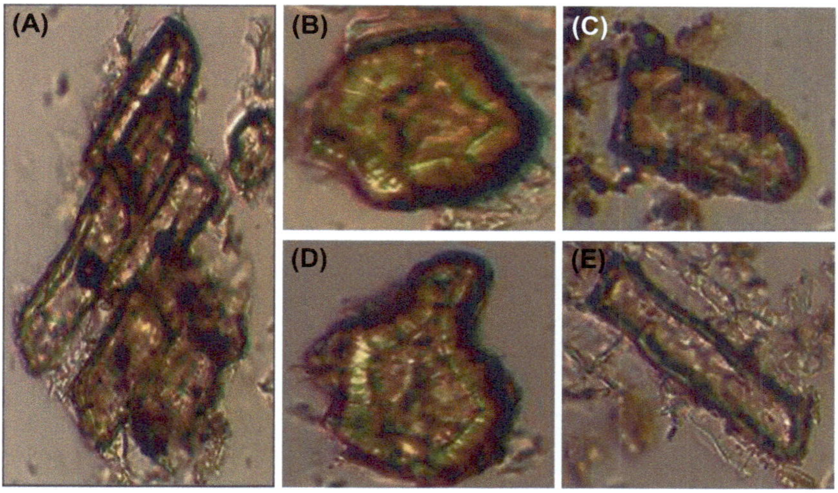

FIG. 5.1.3 (A–E) – Sclereids of various shapes and sizes from *Aegle marmelos* bark powder.

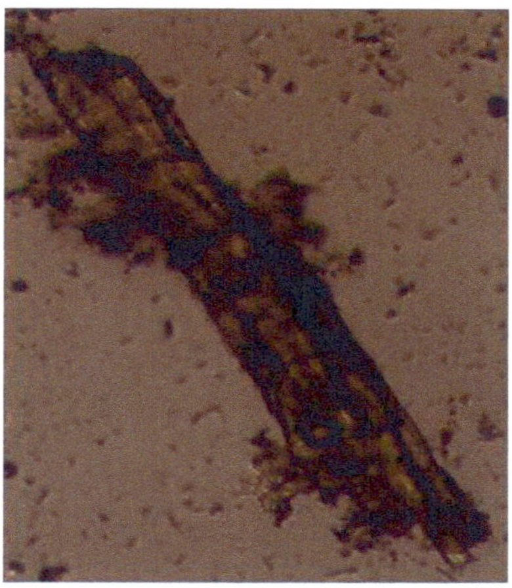

FIG. 5.1.4 Fibres from *Aegle marmelos* bark powder.

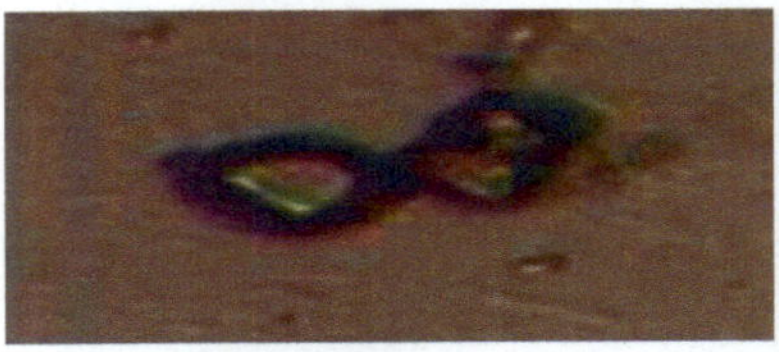

FIG. 5.1.5 Prismatic crystals from *Aegle marmelos* bark powder.

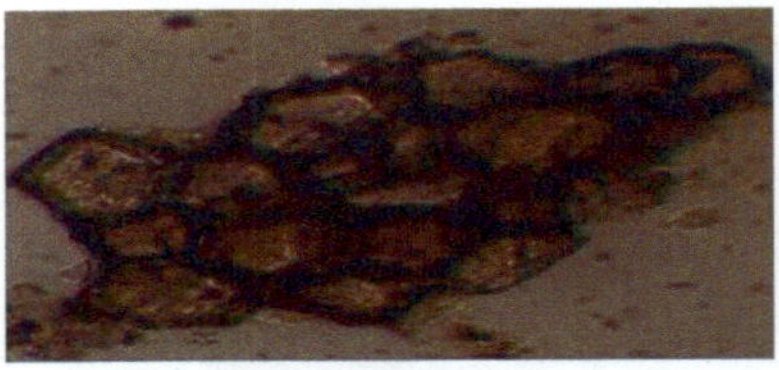

FIG. 5.1.6 Cork cells from *Aegle marmelos* bark powder.

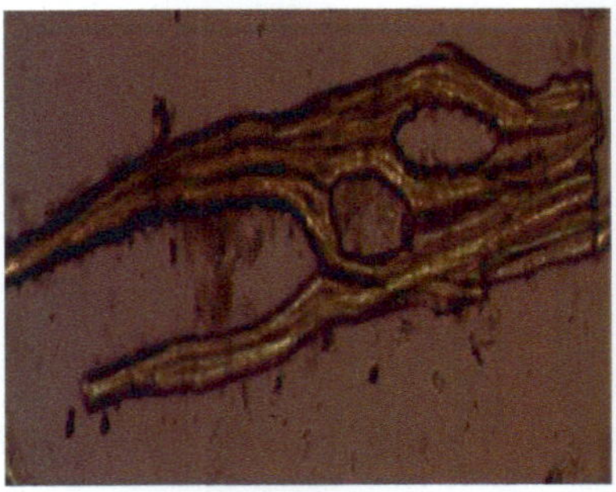

FIG. 5.1.7 Medullary rays from *Aegle marmelos* bark powder.

References

[1] Government of India, Department of AYUSH, Ayurvedic Pharmacopoeia of India. Part I, vol. IV, Ministry of Health and welfare, Ghaziabad, 1999–2017, p. 14.

[2] A.K. Gupta, N. Tandon, M. Sharma, Quality Standards of Indian Medicinal Plants, vol. 8, Medicinal Plant Unit, ICMR, New Delhi, India, 2008, p. 19.

[3] M.G. Chauhan, A.P.G. Pillai, Microscopic Profile of Powdered Drugs Used in Indian System of Medicine, Bark Drugs, vol. 1, Institute of Ayurvedic Medicinal Plant Sciences, Jamnagar, Gujarat, India, 2005, p. 30.

Chapter 5.2

Ailanthus excelsa

Ailanthus excelsa Roxb. Simarubaceae

Common vernacular names: Assamese – Aralu; Gujarati – Aralavo; Hindi – Arlu, Maruk, Ghoda Karanj; Kannada – Hiremara Hebbever; Kashmiri – Merumaram, Mattipongilyam; Malayalam – Merumaram, Mattipongilyam; Marathi – Ghoda Karanj; Oriya – Dakshinakabala, Mahala; Punjabi – Aruo; Tamil – Peruvagai; Telugu – Peddmanu [1] (Fig. 5.2.1).

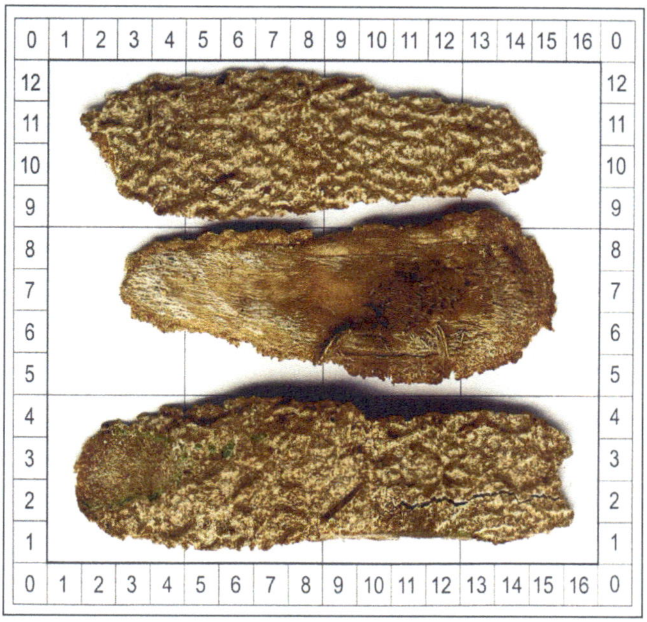

FIG. 5.2.1 *Ailanthus excelsa* bark.

Macroscopic description: Shape – Thick, hard and rough, **Outer bark** – Yellowish grey and rough due to presence of longitudinal ridges, **Inner bark** – Yellowish white and fibrous, **Fracture** – Fibrous, **Odour** – Disagreeable when fresh.

Powdered Crude Drug Microscopy of Leaves and Barks. https://doi.org/10.1016/B978-0-12-818092-1.00006-3

Major chemical constituents: β-Sitosterol, Quassinoids, Atlantic Acid, 2,6-Dimethoxy-Benzoquinone and Melanthin [2].

Traditional uses: Diarrhoea, amoebic dysentery, chronic giardiasis, dyspepsia, abdominal spasm anorectal disease, haemorrhoids [2].

Powder microscopic description: Brownish yellow powder showing abundant sclereids and stone cells, usually in groups of 2–3 with considerable variation in size and shape, some being isodiametric with thick pitted striated walls, while others with three sided thickening, elongated cylindrical, fibre like, with narrow or broad lumen and with distinct pits. The abundant prismatic crystals of calcium oxalate scattered as such. The fragments of cork, the cells were thick walled, polygonal to hexagonal in shape [3]. The general as well as specific features of *A.excelsa* bark powder are illustrated in Table 5.2.1 and Figs. 5.2.2–5.2.5 respectively.

TABLE 5.2.1

	Specific features		General features
	Stone cells	**Sclereids**	**Calcium oxalate crystals**
Occurrence	Isolated	Isolated	Prismatic
	Group	Group	
	Abundant	Abundant	**Cork**
Walls	Thin walled	Thick walled	Polygonal
	Striations	Striations	Pentagonal
	Radiating pits		Hexagonal
Shape	Triangular	Fibre like	
	Oval	Very long	
	Spherical		
Lumen	Pitted	Narrow	
	Wide	Pitted	

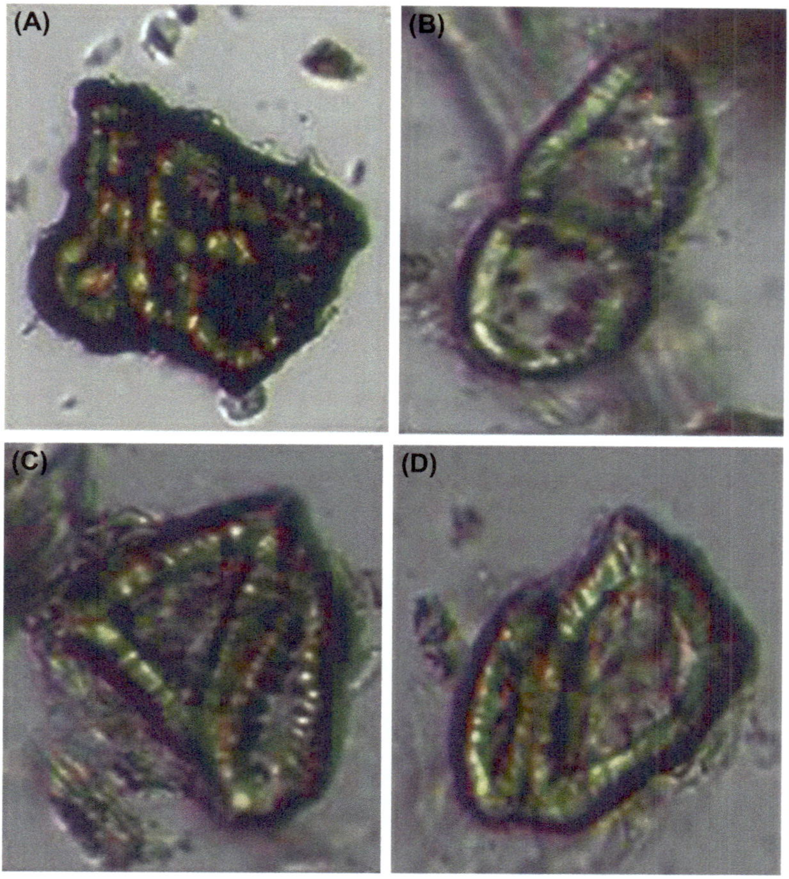

FIG. 5.2.2 (A–D) – Stone cells of various shapes and sizes from *Ailanthus excelsa* bark powder.

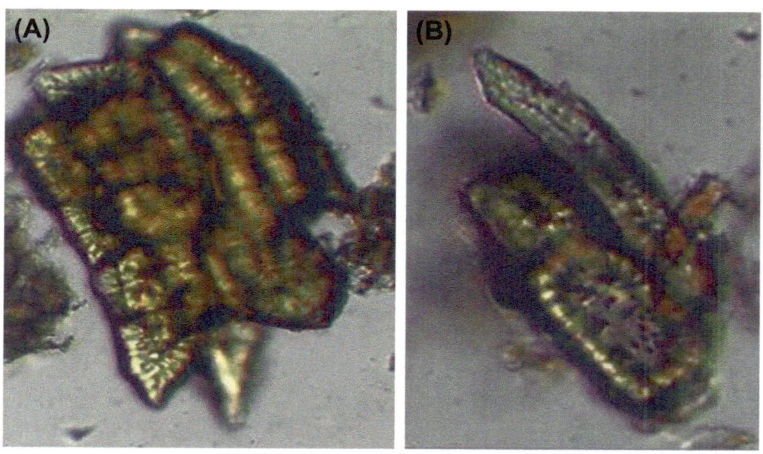

FIG. 5.2.3 (A and B) – Group of sclereids of various shapes and sizes from *Ailanthus excelsa* bark powder.

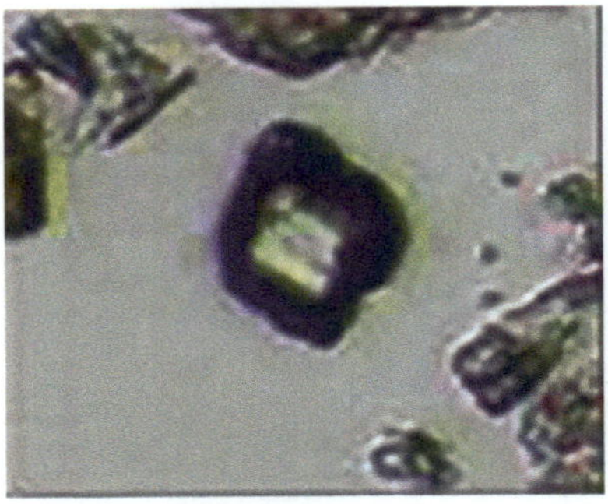

FIG. 5.2.4 Prismatic crystals from *Ailanthus excelsa* bark powder.

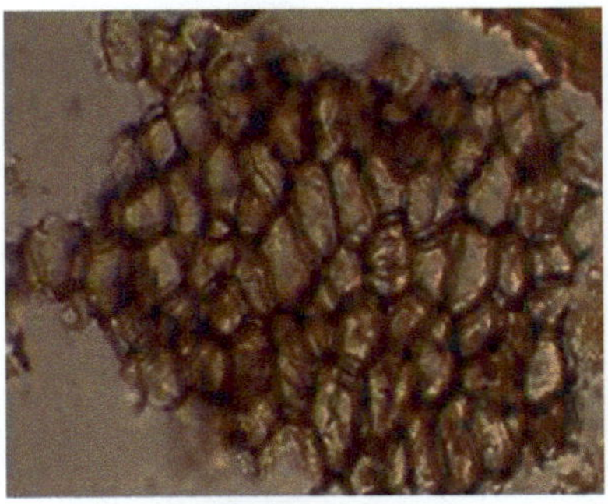

FIG. 5.2.5 Cork cells from *Ailanthus excelsa* bark powder.

References

[1] Government of India, Department of AYUSH, Ayurvedic Pharmacopoeia of India. Part I, vol. III, Ministry of Health and welfare, Ghaziabad, 1999–2017, p. 16.

[2] A.K. Gupta, N. Tandon, M. Sharma, Quality Standards of Indian Medicinal Plants, vol. 3, Medicinal plant unit, ICMR, New Delhi, India, 2008, p. 20.

[3] M.G. Chauhan, A.P.G. Pillai, Microscopic Profile of Powdered Drugs Used in Indian System of Medicine, Bark Drugs, vol. 1, Institute of Ayurvedic Medicinal Plant Sciences, Jamnagar, Gujarat, India, 2005, p. 08.

Chapter 5.3

Alstonia scholaris

Alstonia scholaris (L.) R.Br. Apocynaceae

Common vernacular names: Assamese – Chatiyan; Bengali – Chatin; English – Dita; Gujarati – Saptaparna, Satvana; Hindi – Chhativan, Satawana; Kannada – Maddale, Hale, Eleyalaga; Kashmiri – Kath; Malayalam – Daivaphal, Ezilampala; Marathi – Satveen; Oriya – Chbatiana, Chatiana; Punjabi – Sathi, Satanna; Tamil – Ezilampalai; Telugu – Edakula Ponna [1] (Fig. 5.3.1).

FIG. 5.3.1 *Alstonia scholaris* bark.

Powdered Crude Drug Microscopy of Leaves and Barks. https://doi.org/10.1016/B978-0-12-818092-1.00007-5

Macroscopic description: Shape – Channeled pieces, **Outer bark** – Dark grey to brown, rough and fissured, **Inner bark** – Brownish buff and somewhat striated and indented, **Fracture** – Short and smooth, **Odour** – Characteristic.

Major chemical constituents: Alkaloids (echitamine, ditamine and echitamidine) [2].

Traditional uses: Fever, malaria, troubles in digestion, tumors, ulcers, asthma [2].

Powder microscopic description: Brownish colour powder showing abundant groups of stone cells of various sizes, shapes and thickness with distinct radiating pits and striations. Plenty of sclereids with highly thickened and striated walls of various shapes and sizes, Abundant prismatic crystals along with hexagonal to pentagonal beaded cork cells [3]. The general as well as specific features of *A. scholaris* bark powder are illustrated in Table 5.3.1 and Figs. 5.3.2–5.3.5 respectively.

TABLE 5.3.1 Specific as well as general characters of bark powder of *Alstonia scholaris.*

	Specific features		General features
	Stone cells	**Sclereids**	**Calcium oxalate crystals**
Occurrence	Isolated	Isolated	Prismatic
	Group	Group	
	Abundant	Abundant	**Cork**
Walls	Thick walled	Thin walled	Pentagonal
	Thin walled	Distinct pits	Hexagonal
	Radiating pits		Beaded walls
Shape	Oval	Spherical	Pitted lumen
	Circular	Rectangular	
		Irregular	
Lumen	Slit like	Wide	
	Pitted	Pitted	
	Wide		

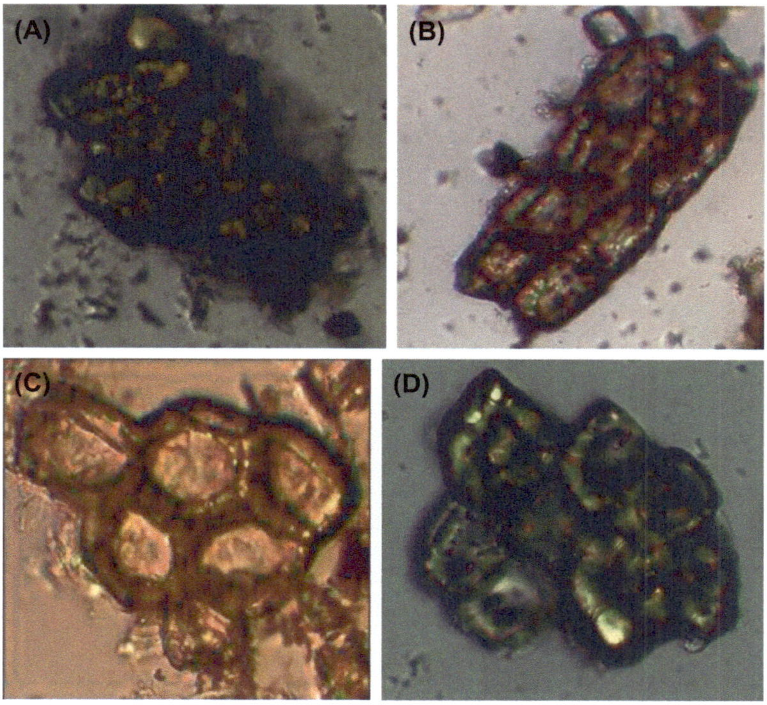

FIG. 5.3.2 (A–D) – Stone cells of various shapes and sizes from *Alstonia scholaris* bark powder.

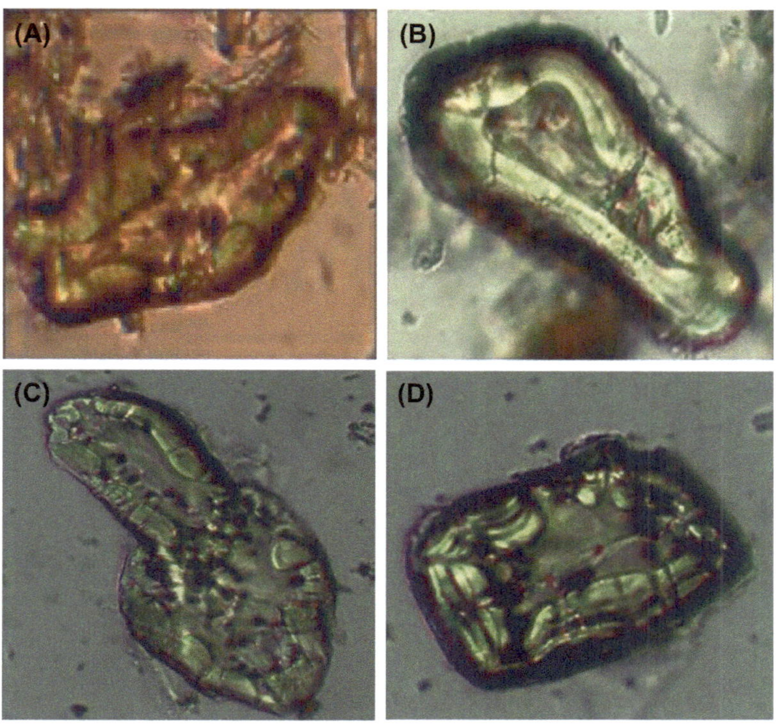

FIG. 5.3.3 (A–D) – Sclereids of various shapes and sizes from *Alstonia scholaris* bark powder.

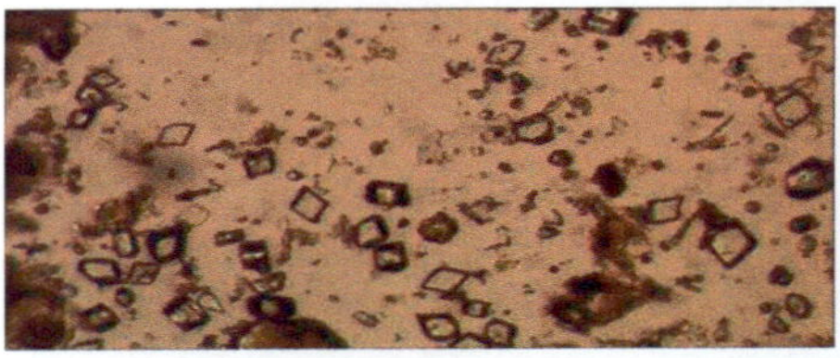

FIG. 5.3.4 Prismatic crystals from *Alstonia scholaris* bark powder.

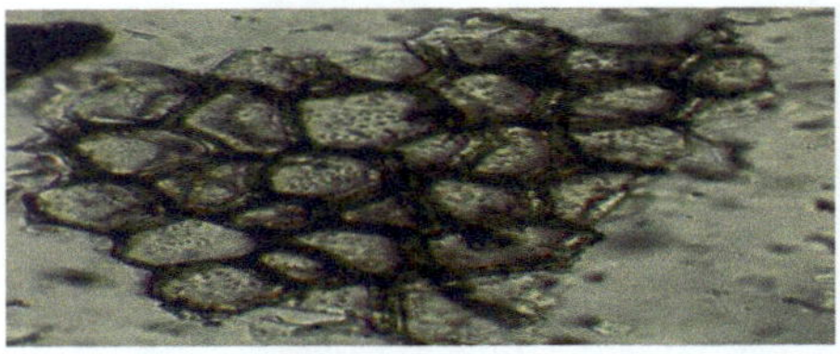

FIG. 5.3.5 Cork cells from *Alstonia scholaris* bark powder.

References

[1] Government of India, Department of AYUSH, Ayurvedic Pharmacopoeia of India. Part I, vol. I, Ministry of Health and Welfare, Ghaziabad, 1999–2017, p. 97.

[2] M.G. Chauhan, A.P.G. Pillai, Microscopic Profile of Powdered Drugs Used in Indian System of Medicine, Bark Drugs, vol. 1, Institute of Ayurvedic Medicinal Plant Sciences, Jamnagar, Gujarat, India, 2005, p. 90.

[3] A.K. Gupta, N. Tandon, M. Sharma, Quality Standards of Indian Medicinal Plants, vol. 3, Medicinal Plant Unit, ICMR, New Delhi, India, 2008, p. 48.

Chapter 5.4

Anogeissus latifolia

Anogeissus latifolia (Roxb. Ex DC.) Wall. ex Guillem. & Perr. *Combretaceae*

Common vernacular names: Bengali – Dhaauyaa gaachh; English – Axle – wood; Gujarati – Dhaavado; Hindi – Baakali, Dhauraa, Dhav, Dhaavaa; Kannada – Dinduge; Malayalam – Vellanava, Malukkanniram; Marathi – Dhaavdaa, Dhaval; Oriya – Dhaau; Tamil – Vellanagai, Vellanamai; Telugu – Chirimaanu [1] (Fig. 5.4.1).

FIG. 5.4.1 *Anogeissus latifolia* bark.

Macroscopic description: Shape – Hard and recurved, **Outer bark** – Pale brown and fairly smooth having small ridges, **Inner bark** – Pale brown and smooth but longitudinally striated, **Fracture** – Clean, **Odour** – Faint.

Major chemical constituents: Phenolic compounds such as ellagic acid, flavellagic acid, quercetin, myricetin and procyandin along with gallotannins, shikimic acid, quinic acid, amino acids, alanine and phenylanine [2].

Powdered Crude Drug Microscopy of Leaves and Barks. https://doi.org/10.1016/B978-0-12-818092-1.00008-7

Traditional uses: Calculus, Piles, Otorrhoea, Leprosy/diseases of skin, Dysuria, Obesity, Anaemia, Metabolic disorder, Disorders of blood, Syphilis/soft chancre, Erysepales [1].

Powder microscopic description: Pale brown powder showing cluster and rosette crystals of calcium oxalate, Oval, spherical to triangular stone cells, Fragments of polygonal to pentagonal cork cells and few sclereids that were irregular in shape. The general as well as specific features of *A. latifolia* bark powder are illustrated in Table 5.4.1 and Figs. 5.4.2–5.4.6 respectively.

TABLE 5.4.1 Specific as well as general characters of bark powder of *Anogeissus latifolia*.

	Specific features		General features
	Stone cells	**Sclereids**	**Calcium oxalate crystals**
Occurrence	Isolated	Few	Cluster
	Group	Group	Rosette
	Abundant		
Walls	Thin walled	Thick walled	**Cork**
	Radiating pits	Distinct Pits	Polygonal
Shape	Triangular	Irregular	Pentagonal
	Oval		Hexagonal
	Spherical		
Lumen	Pitted	Pitted	
	Wide	Narrow	

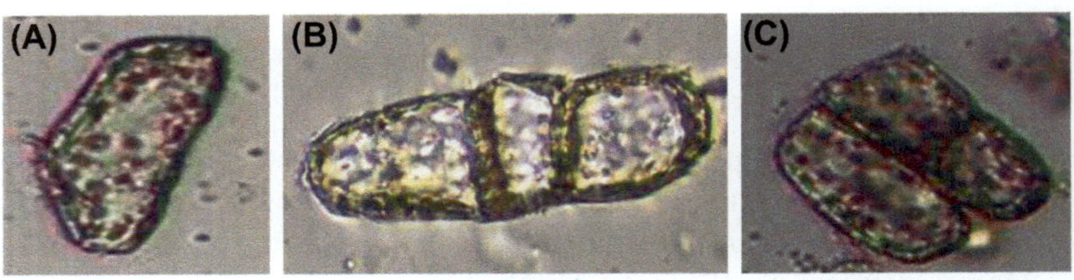

FIG. 5.4.2 (A–C) – Stone cells of various shapes and sizes from *Anogeissus latifolia* bark powder.

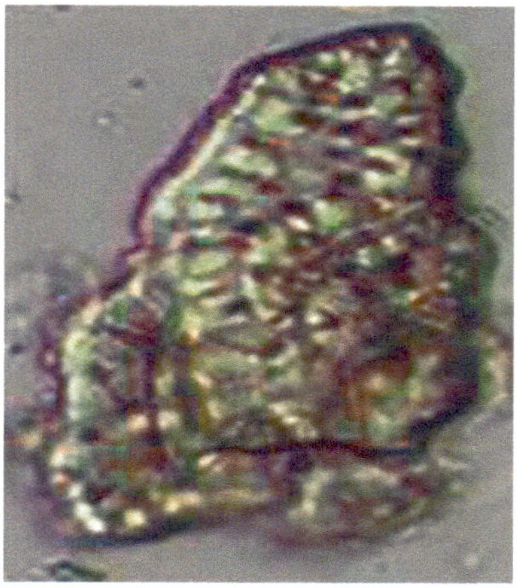

FIG. 5.4.3 Sclereid from *Anogeissus latifolia* bark powder.

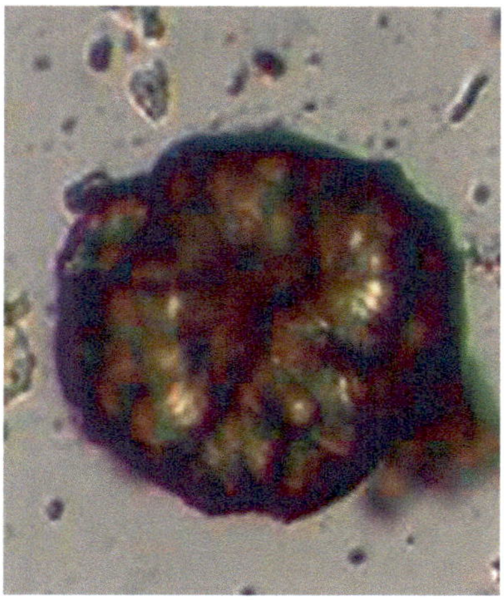

FIG. 5.4.4 Cluster crystal from *Anogeissus latifolia* bark powder.

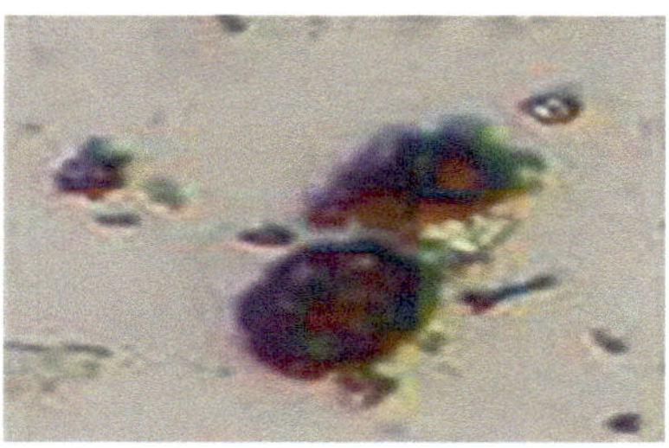

FIG. 5.4.5 Rosette crystal from *Anogeissus latifolia* bark powder.

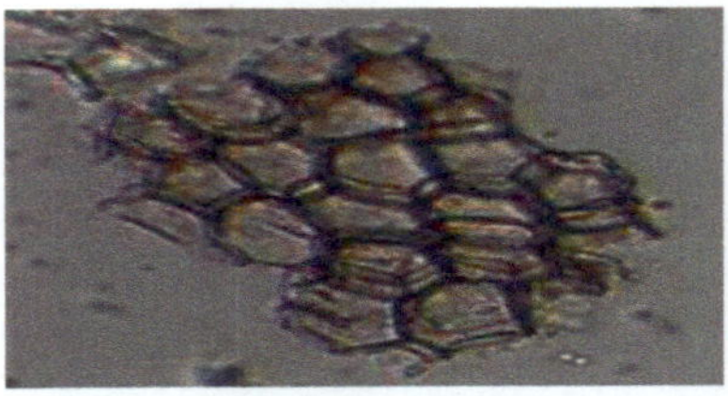

FIG. 5.4.6 Cork cells from *Anogeissus latifolia* bark powder.

References

[1] Government of India, Department of AYUSH, Ayurvedic Pharmacopoeia of India. Part I, vol. VI, Ministry of Health and Welfare, Ghaziabad, 1999–2017, p. LXXV.

[2] A.K. Gupta, N. Tandon, M. Sharma, Quality Standards of Indian Medicinal Plants, vol. 3, Medicinal Plant Unit, ICMR, New Delhi, India, 2008, p. 61.

Chapter 5.5

Artocarpus heterophyllus

Artocarpus heterophyllus Lam. Moraceae

Common vernacular names: Assamese – Kanthal; Bengali – Katal, Kantal, Kathal, Phanas; English – Jack-fruit tree, Indian Jack fruit; Gujarati – Phanus; Hindi – Kathar, Kathal, Katahala; Kannada – Hebba alasu; Malayalam – Chakka; Marathi – Phanasa; Oriya: Panasa, Ponoso; Punjabi – Katahala; Tamil – Pala; Telugu – Panasa; Urdu – Katahal [1] (Fig. 5.5.1).

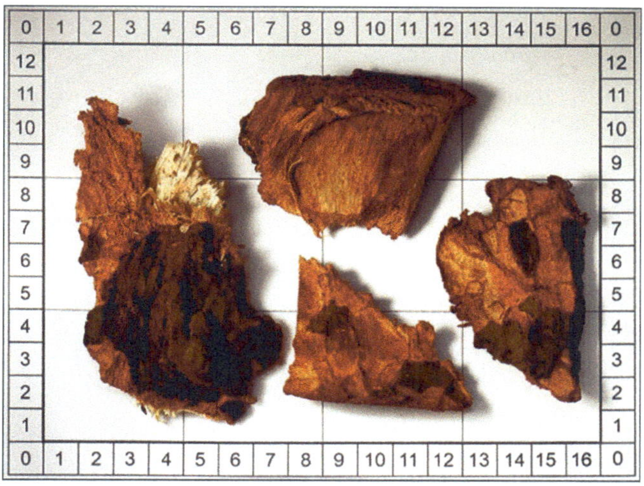

FIG. 5.5.1 *Artocarpus heterophyllus* bark.

Macroscopic description: Shape – Hard and channeled, **Outer bark** – Reddish brown with longitudinal ridges, **Inner bark** – Pale brown and smooth, **Fracture** – Short showing Creamish interior, **Odour** – Indistinct.

Major chemical constituents: β-sitosterol, cycloartenone, cycloartenol; tannins [1].

Powdered Crude Drug Microscopy of Leaves and Barks. https://doi.org/10.1016/B978-0-12-818092-1.00009-9

Traditional uses: Diarrhoea, Burning sensation, Bleeding disorder, Inflammation, Skin diseases [1].

Powder microscopic description: Reddish brown powder showing abundant stone cells and sclereids of all shape and sizes spreaded throughout the field. Overnight soaking of bark powder in water was enough to visualize all the characters but treatment with potassium chlorate enhanced it. Abundant starch grains-simple, compound, spherical, oval appeared on the slide after treatment with iodine solution. Stone cells attached to fibres, fragments of cork cells-polygonal, pentagonal to hexagonal in shape and few prismatic and rosette crystals of calcium oxalate were also seen. The general as well as specific features of *A. heterophyllus* bark powder are illustrated in Table 5.5.1 and Figs. 5.5.2–5.5.8 respectively.

TABLE 5.5.1 Specific as well as general characters of bark powder of *Artocarpus heterophyllus.*

	Specific features		General features
	Stone cells	**Sclereids**	**Calcium oxalate crystals**
Occurrence	Isolated	Isolated	Prismatic
	Group	Group	Cluster
	Abundant	Abundant	**Starch Grains**
Walls	Thick walled	Thick walled	Simple
	Thin walled	Thin walled	Compound
	Striations	Distinct pits	Oval
	Radiating pits		Spherical
Shape	Oval	Oval	**Cork**
	Circular	Rectangular	Hexagonal
	Isodiametric	Pointed ends	Pentagonal
	Rectangular	Pentagonal	Thick walled
	Rounded	Triangular	Polygonal
	Spherical		**Phloem Parenchyma**
Lumen	Wide	Wide	Oval
	Narrow	Pitted	Spherical
	Pitted		

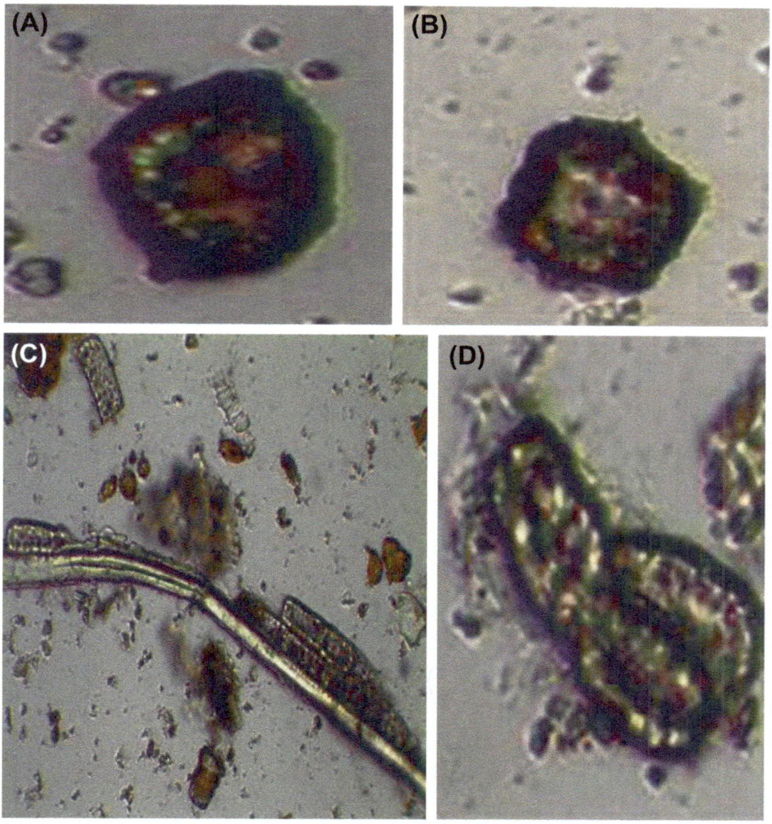

FIG. 5.5.2 (A–D) – Stone cells of various shapes and sizes from *Artocarpus heterophyllus* bark powder.

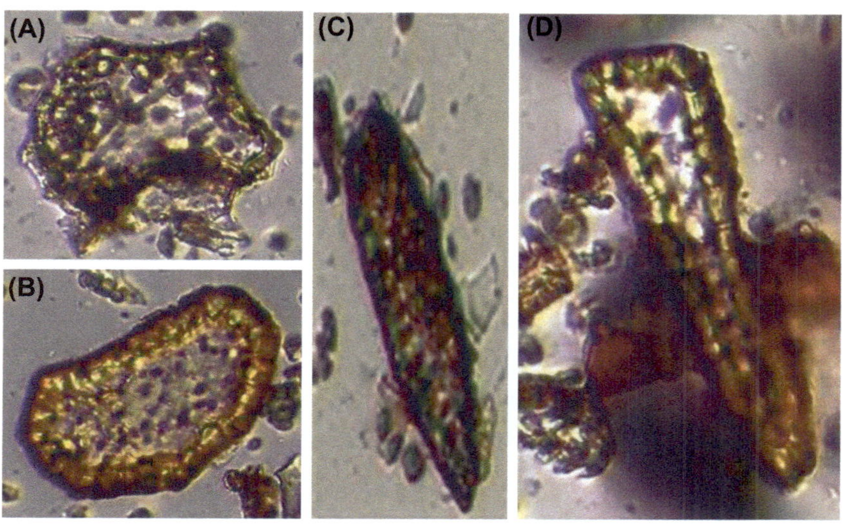

FIG. 5.5.3 (A–D) – Sclereids of various shapes and sizes from *Artocarpus heterophyllus* bark powder.

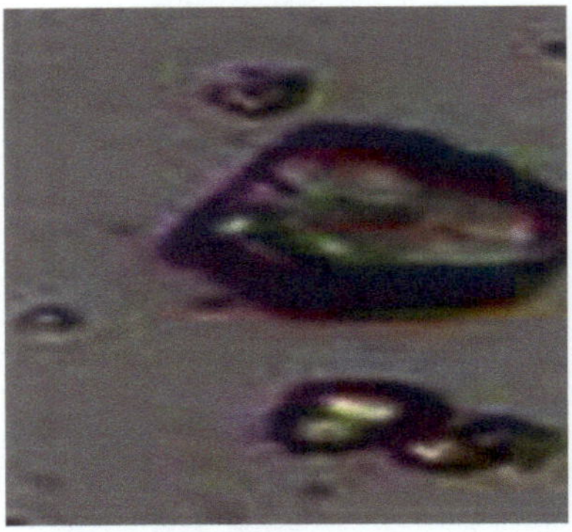

FIG. 5.5.4 Prismatic crystals from *Artocarpus heterophyllus* bark powder.

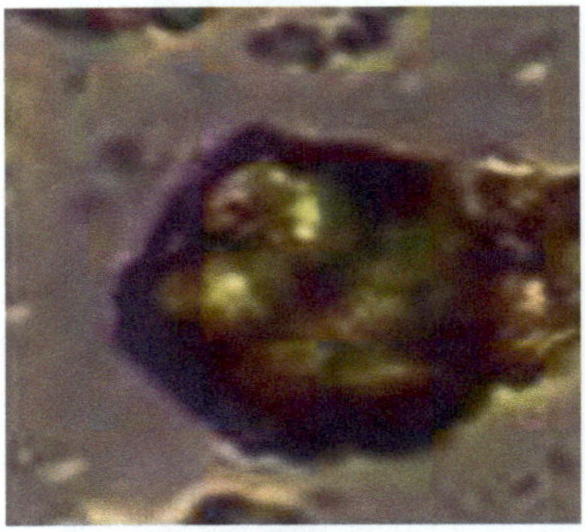

FIG. 5.5.5 Cluster crystals from *Artocarpus heterophyllus* bark powder.

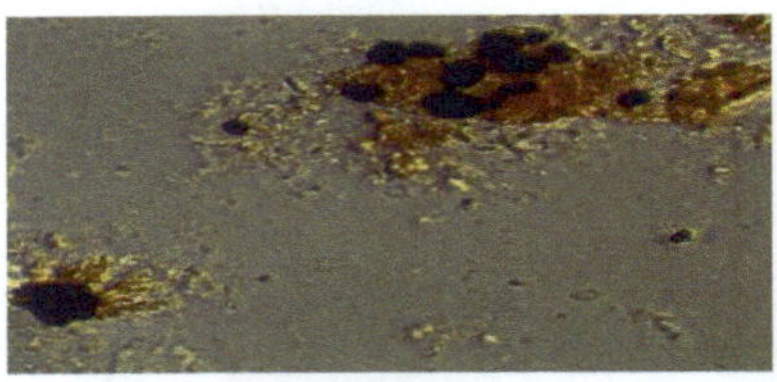

FIG. 5.5.6 Starch grains from *Artocarpus heterophyllus* bark powder.

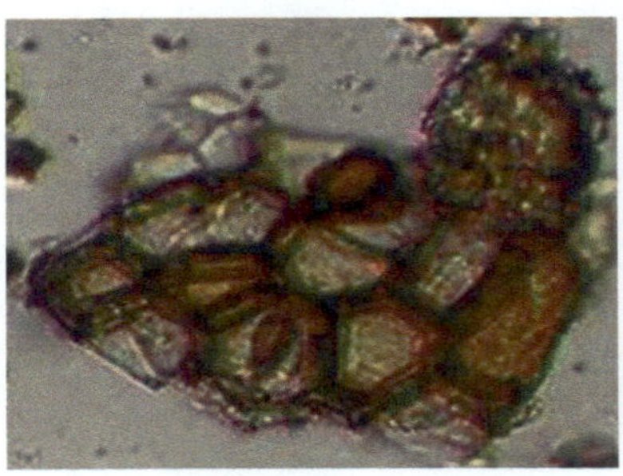

FIG. 5.5.7 Cork cells from *Artocarpus heterophyllus* bark powder.

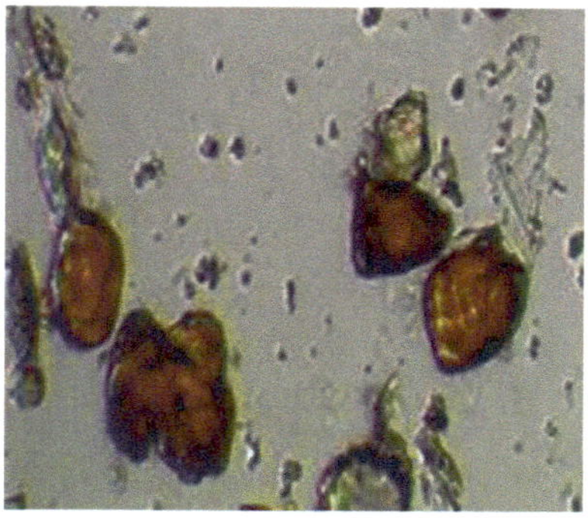

FIG. 5.5.8 Parenchymatous cells from *Artocarpus heterophyllus* bark powder.

Reference

[1] Government of India, Department of AYUSH, Ayurvedic Pharmacopoeia of India. Part I, vol. VI, Ministry of Health and Welfare, Ghaziabad, 1999–2017.

Chapter 5.6

Bombax ceiba

Bombax ceiba L. Malvaceae

Common vernacular names: Assamese – Semul; Bengali – Shimul, Simul; English – Silk-Cotton Tree; Gujarati – Shemalo; Hindi – Semal, Semar; Kannada – Kempuburunga; Malayalam – Mullilavu; Marathi – Sanvar, Katesavar; Punjabi – Simble; Tamil – Elavam; Telugu – Buruga; Urdu – Sembhal [1] (Fig. 5.6.1).

FIG. 5.6.1 *Bombax ceiba* bark.

Macroscopic description: Shape – Curved or channeled, **Outer bark** – Pale ashy to silvery grey, longitudinally wrinkled, furrowed and fissured, **Inner bark** – Brownish, longitudinally striated, **Fracture** – Outer splintery, inner fibrous, **Odour** – Not characteristic.

Major chemical constituents: Lupeol, Shamimicin, β-sitosterol [2].

Traditional uses: Antihaemorrhagic [2].

Powder microscopic description: Brownish powder showing fragments of numerous, very long septate fibres, with highly thickened walls and irregular margin, fragments of cork with polygonal to hexagonal cells, groups of stone

Powdered Crude Drug Microscopy of Leaves and Barks. https://doi.org/10.1016/B978-0-12-818092-1.00010-5

TABLE 5.6.1 Specific as well as general characters of bark powder of *Bombax ceiba*.

	Specific features		General features
	Stone cells	**Fibres**	**Calcium oxalate crystals**
Occurrence	Abundant	Numerous	Rosette
	Isolated	Isolated	
	Group		**Cork**
Walls	Thin walled	Thick walled	Hexagonal
	Radiating pits		Polygonal
	Distinct pits		
	Beaded walls		
Shape	Pentagonal	Very long	
	Oval	Irregular margin	
	Spherical		
Lumen	Pitted	Narrow	
	Wide	Uneven	

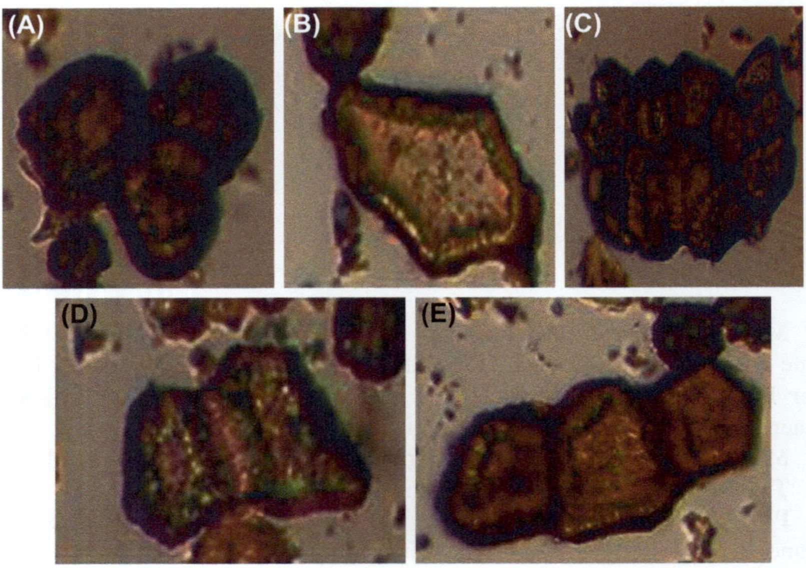

FIG. 5.6.2 (A–E) – Stone cells of various shapes and sizes from *Bombax ceiba* bark powder.

cells – rectangular, oval to spherical in shape, with radiating pitted or beaded walls and wide lumen and abundant rosette and cluster crystals of calcium oxalate [3]. The general as well as specific features of *B. ceiba* bark powder are illustrated in Table 5.6.1 and Figs. 5.6.2–5.6.5 respectively.

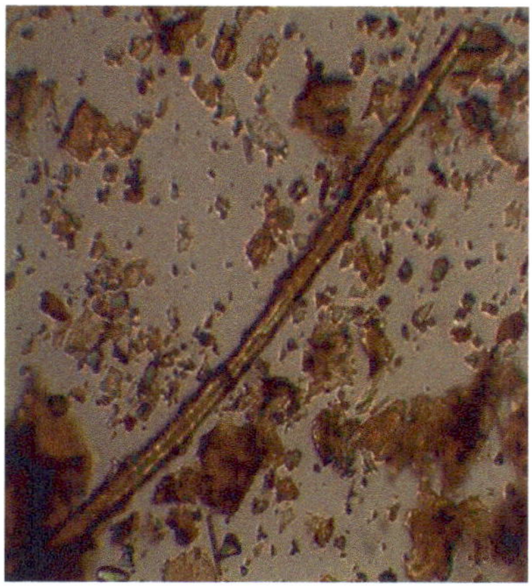

FIG. 5.6.3 Fibres from *Bombax ceiba* bark powder.

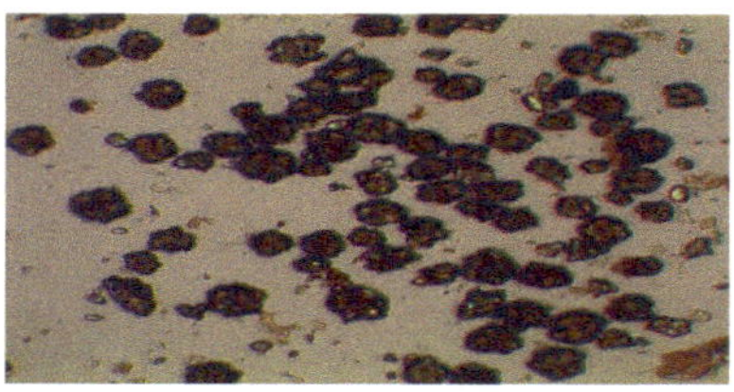

FIG. 5.6.4 Rosette crystals from *Bombax ceiba* bark powder.

FIG. 5.6.5 Cork cells from *Bombax ceiba* bark powder.

References

[1] Government of India, Department of AYUSH, Ayurvedic Pharmacopoeia of India. Part I, vol. III, Ministry of Health and Welfare, Ghaziabad, 1999–2017, p. 185.

[2] A.K. Gupta, N. Tandon, M. Sharma, Quality Standards of Indian Medicinal Plants, vol. 5, Medicinal Plant Unit, ICMR, New Delhi, India, 2008, p. 126.

[3] M.G. Chauhan, A.P.G. Pillai, Microscopic Profile of Powdered Drugs Used in Indian System of Medicine, Bark Drugs, vol. 1, Institute of Ayurvedic Medicinal Plant Sciences, Jamnagar, Gujarat, India, 2005, p. 88.

Chapter 5.7

Buchanania lanzan

Buchanania lanzan Spreng. ***Anacardiaceae***

Common vernacular names: Bengal – Chironji, Pial; English – Calumpang Nut Tree; Gujarati – Charoli; Hindi – Chiraunji, Piyaar, Chironji; Kannada – Kolatmavu, Chalaali; Malayalam – Priyaalam, Mural maram; Marathi – Chaaroli Jhaada; Oriya – Char, Charakoli, Priyal; Punjabi – Chironji; Tamil – Saarapparuppu; Telugu – Sarapappu Chettu, Chinna morilli Mori, Saara; Urdu – Habb-us-Samena [1] (Fig. 5.7.1).

FIG. 5.7.1 *Buchanania lanzan* bark.

Powdered Crude Drug Microscopy of Leaves and Barks. https://doi.org/10.1016/B978-0-12-818092-1.00011-7

Macroscopic description: Shape – Recurved, flat and more or less channeled, **Outer bark** – Greyish brown and rough due to formation of fissures, **Inner bark** – Reddish brown and fibrous, **Fracture** – Fibrous, **Odour** – Characteristic.

Major chemical constituents: Alkaloids, tannins, saponins, reducing sugars, triterpenoids and flavonoids [2].

Traditional uses: Diarrhea and fever [2].

Powder microscopic description: Greyish brown powder showing fragments of cork cells-oval, pentagonal to hexagonal in shape, abundant prismatic crystals of calcium oxalate, isolated and group of stone cells – pitted and wide lumen, thin walled, oval, circular, rectangular to irregular in shape, few stone cells with narrow lumen, rarely occurring sclereids, fragments of medullary rays embedded with dark brown content and simple fibre fragment. The general as well as specific features of *B. lanzan* bark powder are illustrated in Table 5.7.1 and Figs. 5.7.2–5.7.6 respectively.

TABLE 5.7.1 Specific as well as general characters of bark powder of *B. lanzan*.

	Specific features		General features
	Stone cells	**Fibres**	**Calcium oxalate crystals**
Occurrence	Abundant	Few	Prismatic
	Isolated	Isolated	
	Group		**Cork**
Walls	Thin walled	Thick walled	Oval
	Radiating pits	Striations	Hexagonal
Shape	Oval	Very long	Pentagonal
	Circular	Bent	Thick walled
	Rectangular		
	Squarish		**Medullary Rays**
Lumen	Wide	Narrow	Uniseriate
	Pitted		Biseriate

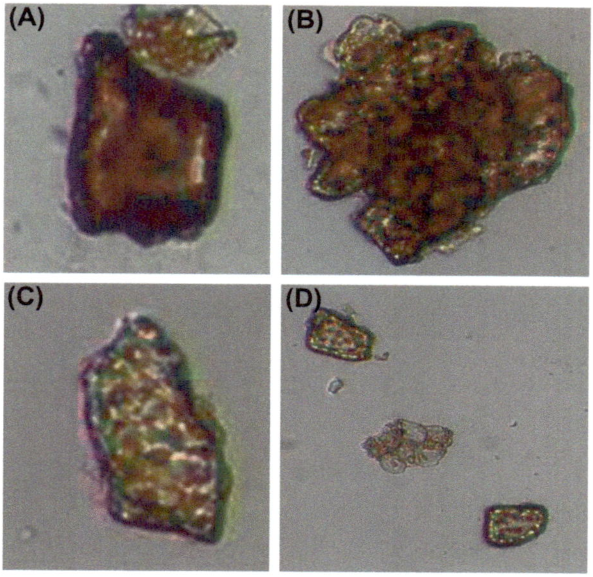

FIG. 5.7.2 (A–D) – Stone cells of various shapes and sizes from *Buchanania lanzan* bark powder.

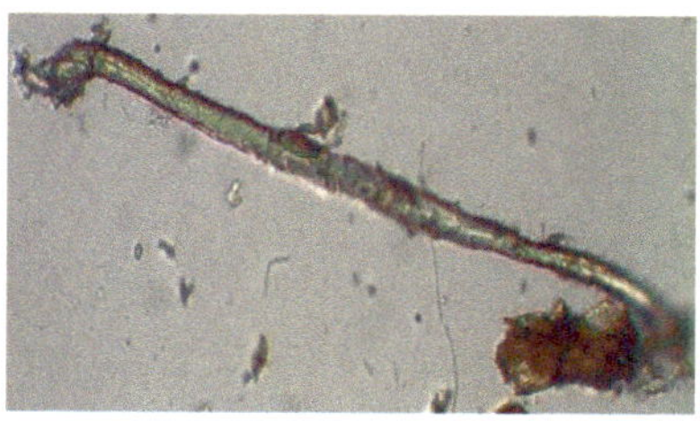

FIG. 5.7.3 Fibre from *Buchanania lanzan* bark powder.

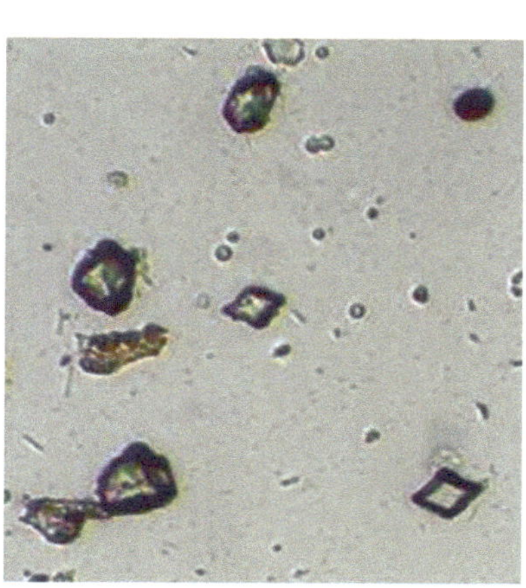

FIG. 5.7.4 Prismatic crystals from *Buchanania lanzan* bark powder.

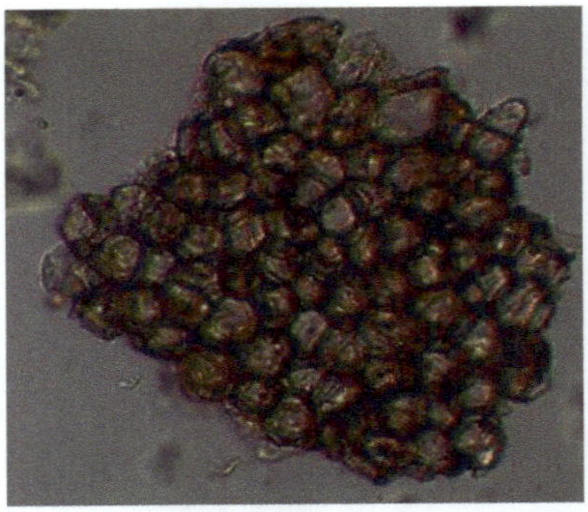

FIG. 5.7.5 Cork cells from *Buchanania lanzan* bark powder.

FIG. 5.7.6 Medullary rays from *Buchanania lanzan* bark powder.

References

[1] Government of India, Department of AYUSH, Ayurvedic Pharmacopoeia of India. Part I, vol. IV, Ministry of Health and Welfare, Ghaziabad, 1999–2017, p. 109.

[2] A.K. Gupta, N. Tandon, M. Sharma, Quality Standards of Indian Medicinal Plants, vol. 12, Medicinal Plant Unit, ICMR, New Delhi, India, 2008, p. 165.

Chapter 5.8

Butea monosperma

Butea monosperma (Lam.) Taub. Fabaceae

Common vernacular names: Bengali – Palash Gachha, Palash, Palas; English – Bastard peak; Gujarati – Kesudo, Khakharo, Khakhapado; Hindi – Dhak, Tesu; Kannada – Muttug, Muttuga, Muttala; Malayalam – Plasu, Camata, Plas, Chama Tha; Marathi – Palas; Punjabi – Palash, Dhak, Tesu; Tamil – Purasu, Paras; Telugu – Moduga, Modugu, Chettu; Urdu – Dhak, Palaspapda [1] (Fig. 5.8.1).

FIG. 5.8.1 *Butea monosperma* bark.

Macroscopic description: Shape – Curved and rough, **Outer bark** – Greyish brown, thick usually peels off exposing light brown surface, presence of shallow and longitudinal fissures, **Inner bark** – Rough and pale brown,

Powdered Crude Drug Microscopy of Leaves and Barks. https://doi.org/10.1016/B978-0-12-818092-1.00012-9

Fracture – Laminated in outer part and fibrous in inner part, **Odour** – Characteristic.

Major chemical constituents: Kinnotannic acid, gallic acid and major glycosides such as butrin [1].

Traditional uses: Hemorrhoids, bowel disorders, wound healing, treatment of worms [1].

Powder microscopic description: Greyish brown powder showing abundant fragments of parenchymatous tissue consisting of rectangular, oval or rounded cells filled with dark orange brown gummy masses, abundant fragments of cork cells-hexagonal to polygonal in shape with beaded walls and pitted lumen, abundant prismatic crystals of calcium oxalate, few starch grains-oval to round in shape and isolated or groups spherical stone cells with pitted, wide lumen and beaded walls, somewhat rectangular in shape and with narrow lumen [2]. The general as well as specific features of *B. monosperma* bark powder are illustrated in Table 5.8.1 and Figs. 5.8.2–5.8.5 respectively.

TABLE 5.8.1 Specific as well as general characters of bark powder of *B. monosperma.*

Specific features		General features
	Stone cells	**Calcium oxalate crystals**
Occurrence	Abundant	Prismatic
	Isolated	
	Group	**Cork**
Walls	Thin walled	Hexagonal
	Beaded	Polygonal
Shape	Rectangular	Pitted lumen
	Triangular	
Lumen	Wide	**Masses**
	Pitted	Orange brown gummy masses

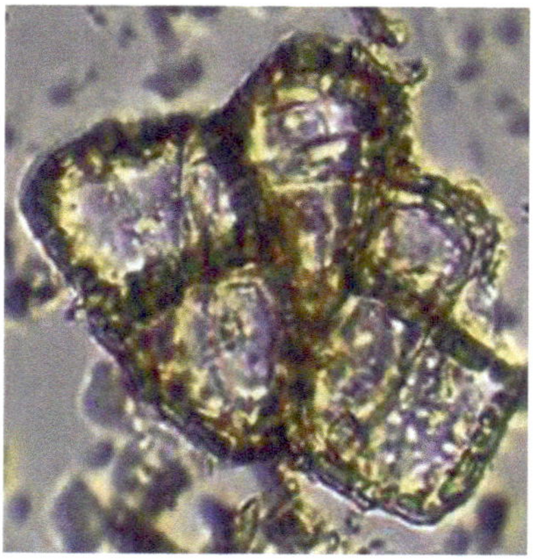

FIG. 5.8.2 Stone cells from *Butea monosperma* bark powder.

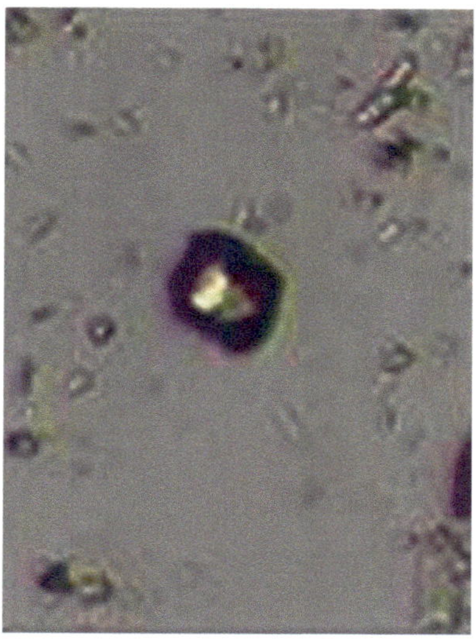

FIG. 5.8.3 Prismatic crystals from *Butea monosperma* bark powder.

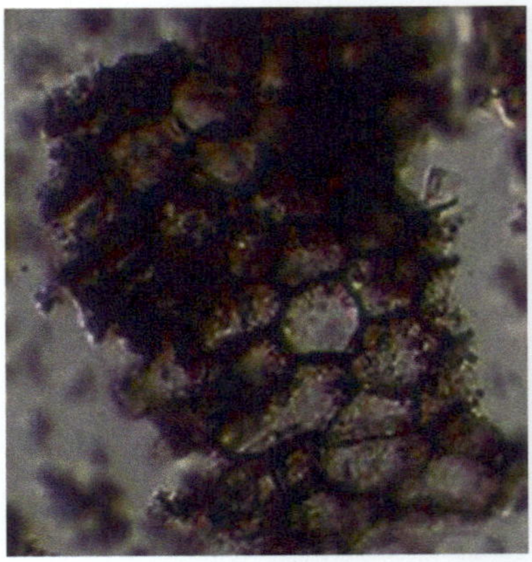

FIG. 5.8.4 Cork cells from *Butea monosperma* bark powder.

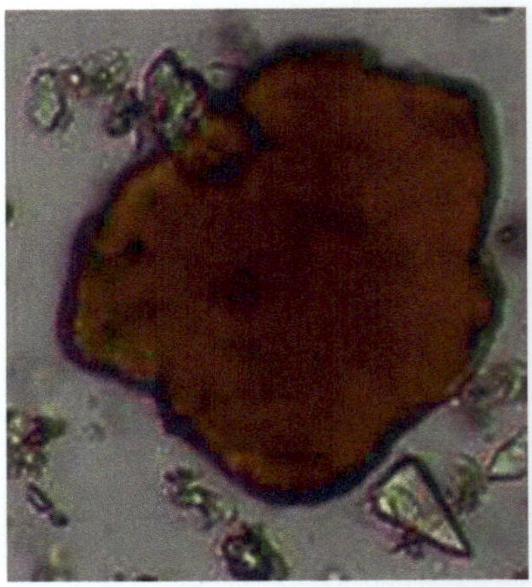

FIG. 5.8.5 Orange brown gummy mass from *Butea monosperma* bark powder.

References

[1] Government of India, Department of AYUSH, Ayurvedic Pharmacopoeia of India. Part I, vol. II, Ministry of Health and Welfare, Ghaziabad, 1999–2017, p. 137.

[2] M.G. Chauhan, A.P.G. Pillai, Microscopic Profile of Powdered Drugs Used in Indian System of Medicine, Bark Drugs, vol. 1, Institute of Ayurvedic Medicinal Plant Sciences, Jamnagar, Gujarat, India, 2005, p. 76.

Chapter 5.9

Calotropis procera

Calotropis procera (Aiton) Dryand. Asclepiadaceae

Common vernacular names: Assamese – Akand, Akan; Bengali – Akanda, Akone; English – Maddar; Gujarati – Aakado; Hindi – Aak, Madar, Akavana; Kannada – Ekka, Ekkagida; Malayalam – Errikku; Marathi – Rui; Oriya – Arakku; Punjabi – Akk; Tamil – Vellerukku, Erukku; Telugu – Jilledu [1] (Fig. 5.9.1).

FIG. 5.9.1 *Calotropis procera* bark.

Macroscopic description: Shape – Channeled, quilled and fibrous pieces, **Outer bark** – Yellowish brown having longitudinal cracks, **Inner bark** – Greenish, smooth with wood tissue attached, **Fracture** – Fibrous, **Odour** – Not distinct.

Powdered Crude Drug Microscopy of Leaves and Barks. https://doi.org/10.1016/B978-0-12-818092-1.00013-0

Major chemical constituents: α - and β- Calotropeols, β-Amyrin, Giganteol, a Colourless wax, small amount of Tetracyclic Terpenes and Traces of Sterols [1].

Traditional uses: Skin diseases, Intestinal worms and Leprosy [1].

Powder microscopic description: Light yellowish green powder showing fragments of cork with thin walled cells and appeared reddish brown in colour, Stone cells occurring in groups with rectangular to elongated shape, few sclereids spherical in shape, phloem fibres appearing in group and individual fibre, thick and can appear either entire or in fragments and simple and round starch grains. The general as well as specific features of *C.procera* bark powder are illustrated in Table 5.9.1 and Fig. 5.9.2A-D, 5.9.3, 5.9.4, 5.9.5 and 5.9.6 respectively.

TABLE 5.9.1 Specific as well as general characters of bark powder of *C.procera.*

	Specific features			General features
	Stone cells	**Sclereids**	**Fibres**	**Cork**
Occurrence	Isolated	Few	Numerous	Polygonal
	Group	Group	Group	Hexagonal
	Abundant			Thin walled
Walls	Thin walled	Thin walled	Faint transverse septa	
	Radiating pits	Distinct pits	Smoothly striated	
Shape	Oval	Spherical	Rectangular	
	Circular		Very long	
Lumen	Wide	Wide	Highly pitted wide lumen	
	Pitted	Pitted		

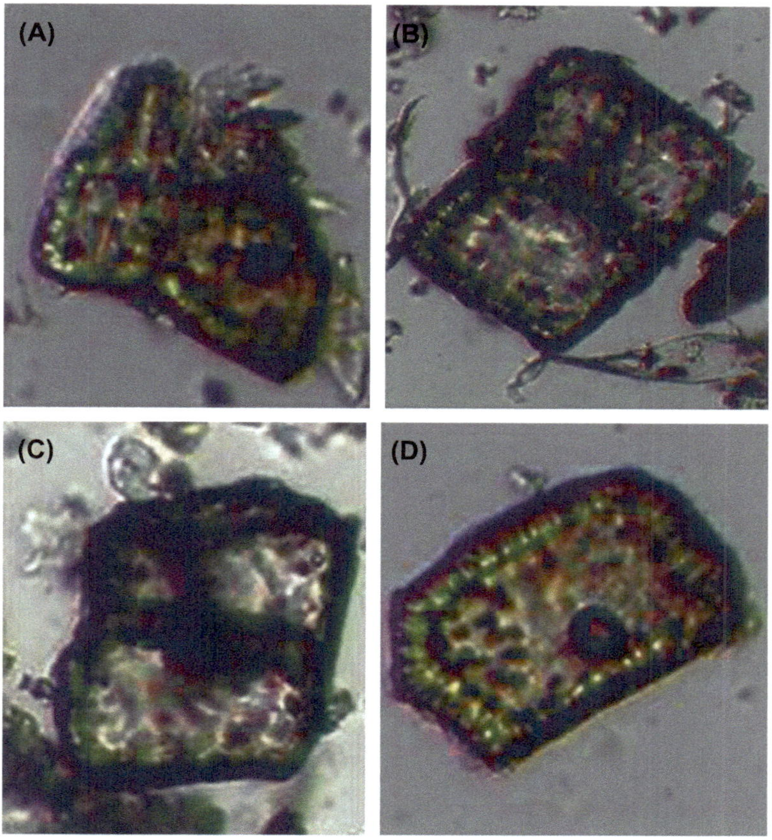

FIG. 5.9.2 (A–D) – Stone cells from *Calotropis procera* bark powder.

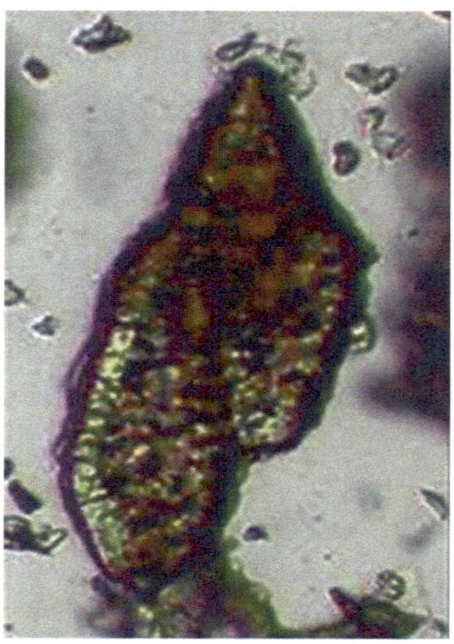

FIG. 5.9.3 Sclereids from *Calotropis procera* bark powder.

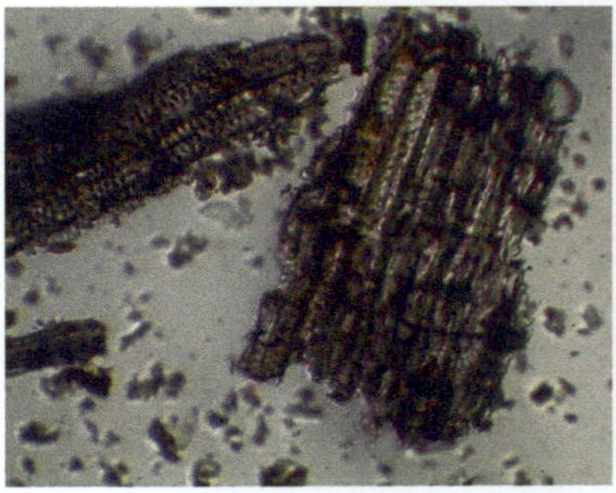

FIG. 5.9.4 Phloem fibres from *Calotropis procera* bark powder.

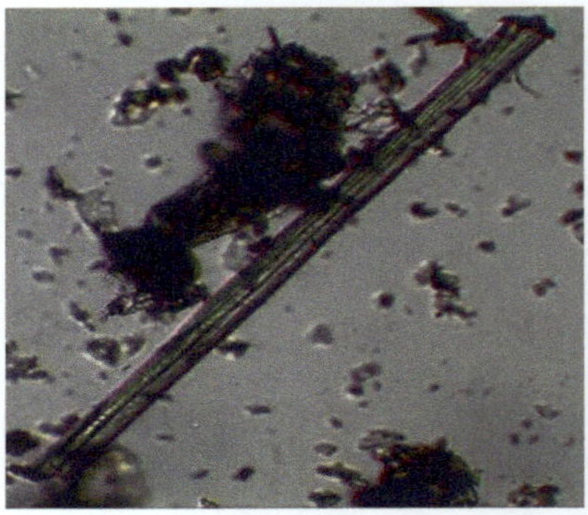

FIG. 5.9.5 Fibre fragment from *Calotropis procera* bark powder.

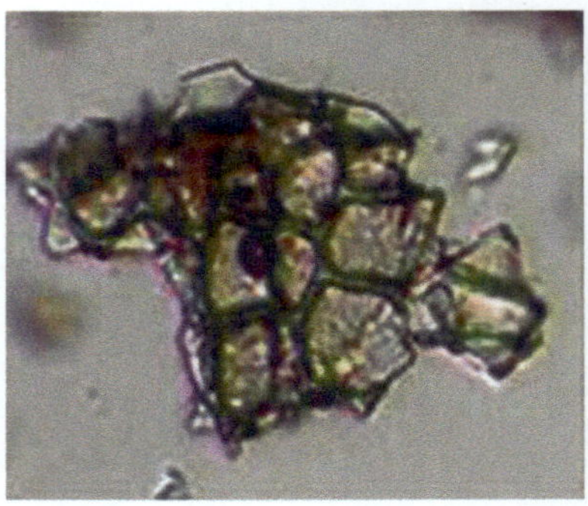

FIG. 5.9.6 Cork cells from *Calotropis procera* bark powder.

Reference

[1] Government of India, Department of AYUSH, Ayurvedic Pharmacopoeia of India. Part I, vol. III, Ministry of Health and welfare, Ghaziabad, 1999-2017, p. 18.

Chapter 5.10

Carissa carandas

Carissa carandas L. Apocynaceae

Common vernacular names: Bengali – Karamach; Gujarati – Karamadan; Hindi – Karijige; Kannada – Karimkar; Malayalam – Karimkar; Marathi – Karamanda; Punjabi – Garna; Tamil – Kalakke [1] (Fig. 5.10.1).

FIG. 5.10.1 *Carissa carandas* bark.

Macroscopic description: Shape – Flat or slightly curved pieces, **Outer bark** – Brownish grey and rough due to longitudinal striations, **Inner bark** – Light yellow and smooth and light in weight, **Fracture** – Short, **Odour** – Characteristic.

Major chemical constituents: β-sitosterol, lupeol, glucosides of odoroside-H, ursolic acid [1].

Traditional uses: Asthma and skin diseases [1].

Powdered Crude Drug Microscopy of Leaves and Barks. https://doi.org/10.1016/B978-0-12-818092-1.00014-2

Powder microscopic description: Greyish brown powder showing abundant group of fibres – thick walled, pointed with slightly uneven margin and often with few obliquely running pits; cork cells – thick walled, polygonal to hexagonal in shape; Plenty of prismatic crystals of calcium oxalate scattered as such throughout the powder; Stone cells, isolated or in groups, isodiametric and with wide lumen with pits; Sclereids few, triangular to pentagonal in shape with wide pitted lumen; abundant fragments of tangentially and radially longitudinally cut medullary rays and starch grains, simple, oval to spherical in shape scattered as such [2]. The general as well as specific features of *C. carandas* bark powder are illustrated in Table 5.10.1 and Fig. 5.10.2–5.10.6 respectively.

TABLE 5.10.1 Specific as well as general characters of bark powder of *C. carandas*.

	Specific features			General features
	Stone cells	**Sclereids**	**Fibres**	**Cork**
Occurrence	Abundant	Few	Numerous	Thick walled
	Group	Group	Group	Pentagonal
Walls	Thick walled	Thin walled	Thick walled	
		Striations	Oblique pits	**Calcium oxalate crystals**
Shape	Polygonal	Polygonal	Irregular margin	Prismatic
	Pentagonal	Pentagonal		
Lumen	Slit like	Wide	Narrow	
		Pitted		

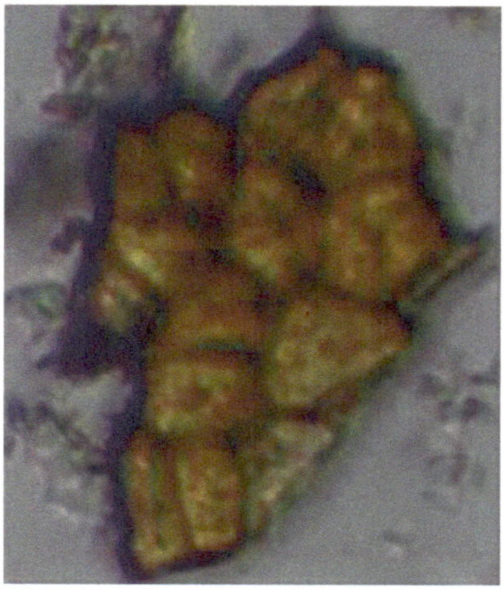

FIG. 5.10.2 Stone cells from *Carissa carandas* bark powder.

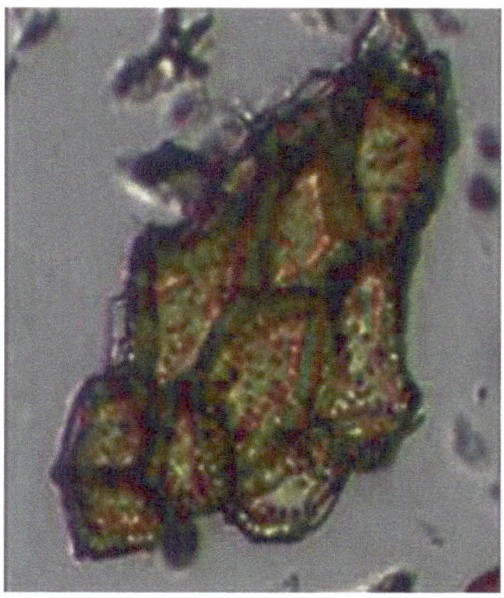

FIG. 5.10.3 Sclereids from *Carissa carandas* bark powder.

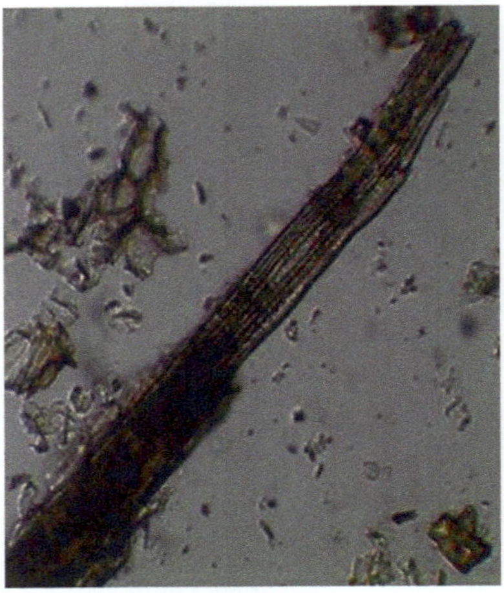

FIG. 5.10.4 Fibres from *Carissa carandas* bark powder.

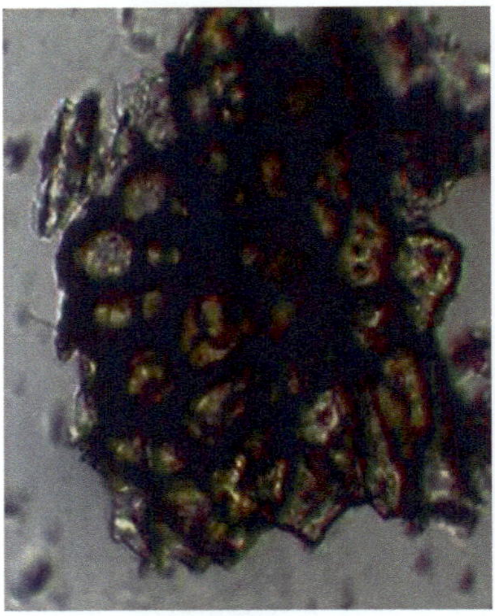

FIG. 5.10.5 Cork cells from *Carissa carandas* bark powder.

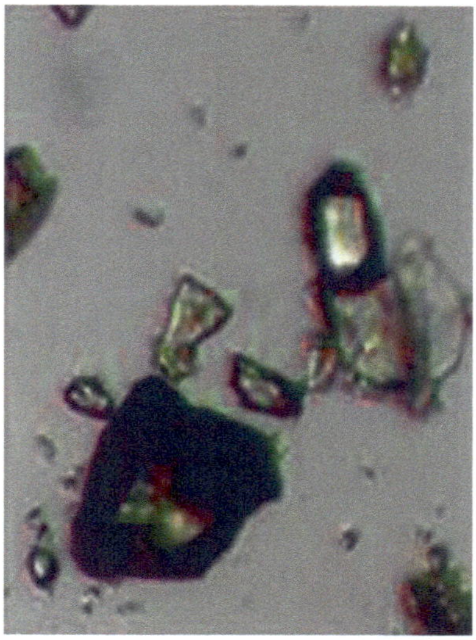

FIG. 5.10.6 Prismatic crystals from *Carissa carandas* bark powder.

References

[1] Government of India, Department of AYUSH, Ayurvedic Pharmacopoeia of India. Part I, vol. II, Ministry of Health and welfare, Ghaziabad, 1999-2017, p. 78.

[2] M.G. Chauhan, A.P.G. Pillai, Microscopic Profile of Powdered Drugs Used in Indian System of Medicine, Bark Drugs, vol. 1, Institute of Ayurvedic Medicinal Plant Sciences, Jamnagar, Gujarat, India, 2005, p. 52.

Chapter 5.11

Cedrela toona

Cedrela toona Roxb. Ex Rottler Meliaceae

Common vernacular names: Bengali – Toongaachha; English – Toon, Red ceder; Gujrati – Toonee; Hindi – Tun, Toonee, Tuni; Kannada – Mandurike, Kempu Gandagheri; Malayalam – Madagirivempu, Ikana, Patukarana; Marathi – Toonee, Kurak; Tamil – Karamusuli, Shevagil Malavembu; Telugu – Nandichettu, Galimanu [1] (Fig. 5.11.1).

Macroscopic description: Shape – Long pieces and channeled, **Outer bark** – Dark grey, rough and rugged due to exfoliation and vertical cracks, **Inner bark** – Red, laminated and fibrous, **Fracture** – Fibrous and splintery, **Odour** – Very mild and pleasant.

Major chemical constituents: Triterpenoids [1].

Traditional uses: Astringent, febrifuge, tonic, antiperiodic, chronic dysentery and wounds [1].

Powder microscopic description: Reddish brown powder showing few fragments of fibres, thin walled with narrow lumen; thick walled cork cells pentagonal to hexagonal in shape; Plenty of prismatic crystals of calcium oxalate scattered as such throughout the powder; Few sclereids with fibre like appearance; abundant fragments of tangentially and radially longitudinally cut

FIG. 5.11.1 *Cedrela toona* bark.

Powdered Crude Drug Microscopy of Leaves and Barks. https://doi.org/10.1016/B978-0-12-818092-1.00015-4

medullary rays and few simple and compound starch grains. The general as well as specific features of *C. toona* bark powder are illustrated in Table 5.11.1 and Fig. 5.11.2–5.11.7 respectively.

TABLE 5.11.1 Specific as well as general characters of bark powder of *C. toona*.

	Specific features		General features	
	Sclereids	**Fibres**	**Medullary rays**	**Calcium oxalate crystals**
Occurrence	Few	Few	Uniseriate	Prismatic
	Isolated	Isolated	Biseriate	
Walls	Thick walled	Thin walled	Tangentially cut	**Starch grains**
	Radiating pits	Striations		Few
	Striations		**Cork**	Simple
Shape	Spherical	Irregular margin	Thick walled	Oval
	Fibre like	Bent	Pentagonal	Compound
Lumen	Narrow	Wide	Hexagonal	Slit hilum

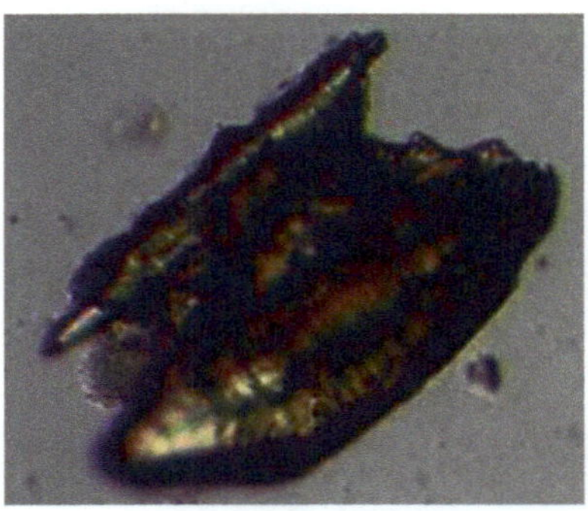

FIG. 5.11.2 Sclereids from *Cedrela toona* bark powder.

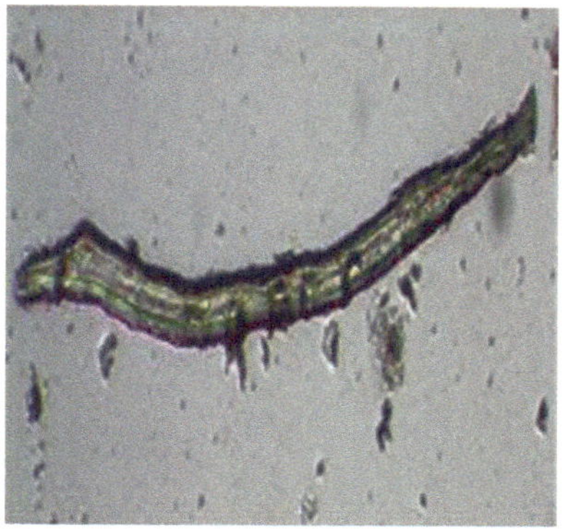

FIG. 5.11.3 Fibre fragment from *Cedrela toona* bark powder.

FIG. 5.11.4 Medullary rays from *Cedrela toona* bark powder.

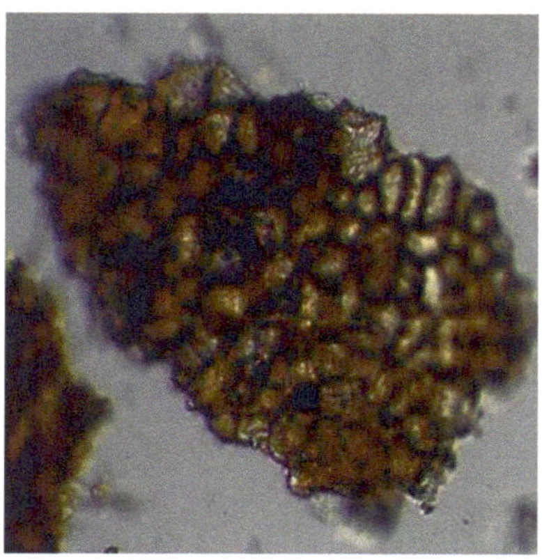

FIG. 5.11.5 Cork cells from *Cedrela toona* bark powder.

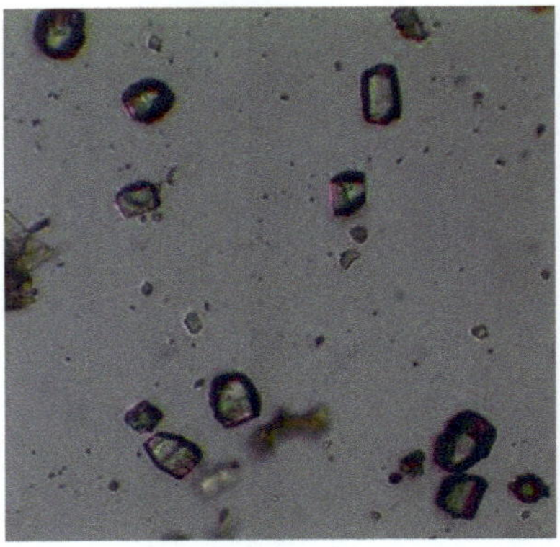

FIG. 5.11.6 Prismatic crystals from *Cedrela toona* bark powder.

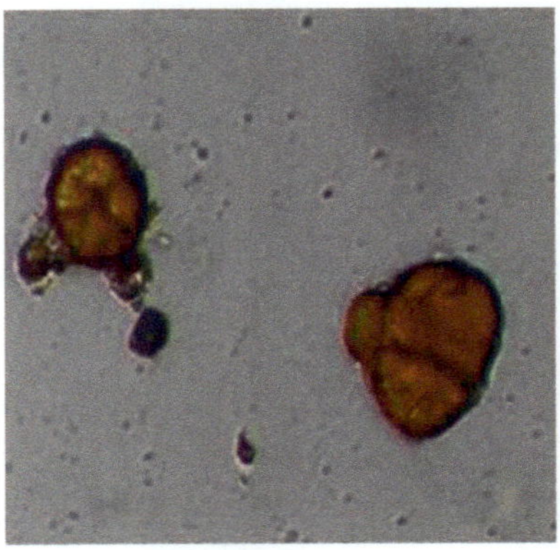

FIG. 5.11.7 Starch grains from *Cedrela toona* bark powder.

Reference

[1] Government of India, Department of AYUSH, Ayurvedic Pharmacopoeia of India. Part I, vol. V, Ministry of Health and welfare, Ghaziabad, 1999-2017, p. 211.

Chapter 5.12

Crataeva nurvula

Crataeva nurvula Buch. -Ham. Capparaceae

Common vernacular names: Bengali – Varuna; English - Three leaved caper; Gujrati – Vayvarno, Varano; Hindi – Baruna, Barna; Kannada – Bipatri, Mattamavu, Neervalamara; Kashmiri – Kath; Malayalam – Neer-matalam; Marathi – Vayavarna, Haravarna, Varon; Oriya – Baryno; Punjabi – Barna, Barnahi; Tamil – Maralingam; Telugu – Bilvarani [1] (Fig. 5.12.1).

FIG. 5.12.1 *Crataeva nurvula* bark.

Macroscopic description: Shape – Tough, channeled, flat or slightly curved; **Outer bark** – Greyish brown, rough and minutely fissured; **Inner bark** – Smooth, creamy and white in colour; **Fracture** – Short and splintery; **Odour** – Not characteristic.

Major chemical constituents: Lupeol, betulinic acid, ceryl alcohol, cadabicine and its diacetate, β-sitosterol, diosgenin, friedelin [2].

Traditional uses: Urolithiasis and dysuria [2].

Powdered Crude Drug Microscopy of Leaves and Barks. https://doi.org/10.1016/B978-0-12-818092-1.00016-6

Powder microscopic description: Greyish brown powder showing abundant stone cells and sclereids of various sizes, shape and thickness, usually with very narrow lumen and highly pitted and striated walls; prismatic crystals scattered as such; fragments of thick walled, polygonal to oval shaped cork cell and fragments of thick walled, pitted parenchymatous cells of the outer phloem with thick striated walls. The general as well as specific features of *C. nurvula* bark powder are illustrated in Table 5.12.1 and Fig. 5.12.2–5.12.6 respectively.

TABLE 5.12.1 Specific as well as general characters of bark powder of *C. nurvula.*

	Specific features		General features
	Stone cells	**Sclereids**	**Calcium oxalate crystals**
Occurrence	Isolated	Isolated	Prismatic
	Group	Group	**Cork**
	Abundant	Abundant	Thick walled
Walls	Thick walled	Thick walled	Hexagonal
	Striations	Striations	Pentagonal
Shape	Oval	Oval	Polygonal
		Spherical	Oval
		Irregular	**Parenchyma**
Lumen	Narrow	Narrow	Pitted

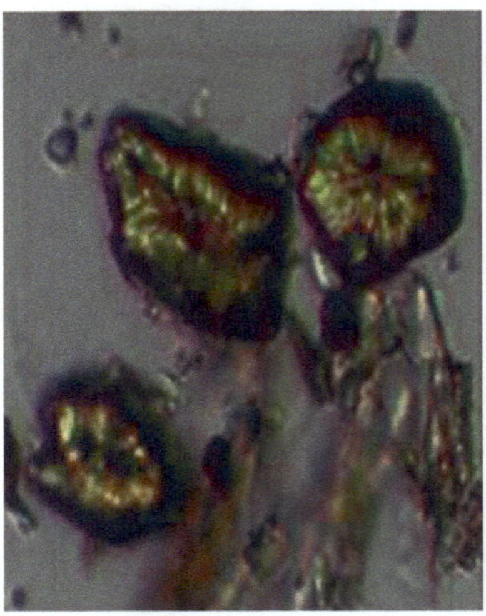

FIG. 5.12.2 Stone cells from *Crataeva nurvula* bark powder.

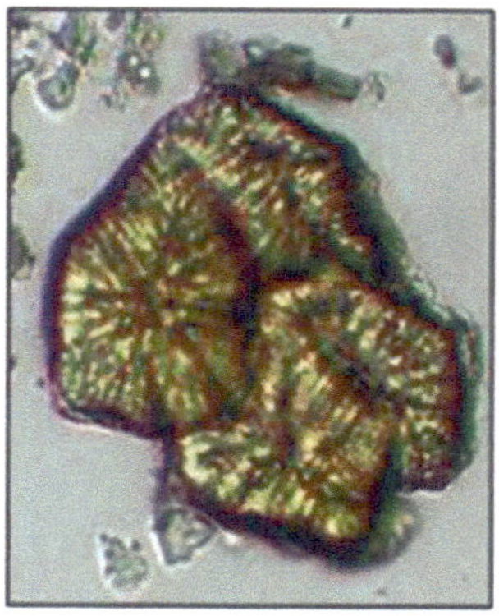

FIG. 5.12.3 Sclereids from *Crataeva nurvula* bark powder.

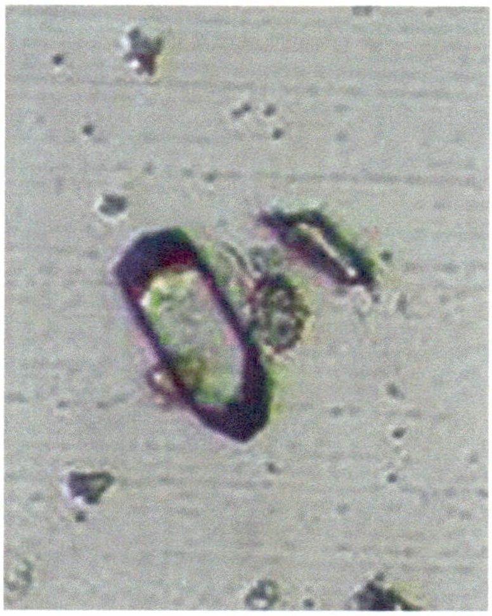

FIG. 5.12.4 Prismatic crystals from *Crataeva nurvula* bark powder.

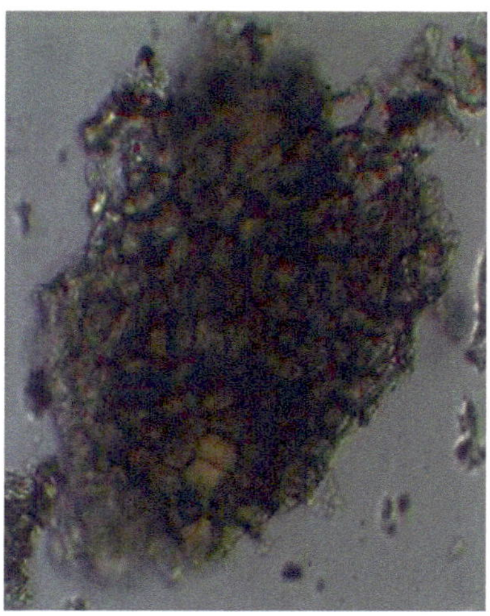

FIG. 5.12.5 Cork cells from *Crataeva nurvula* bark powder.

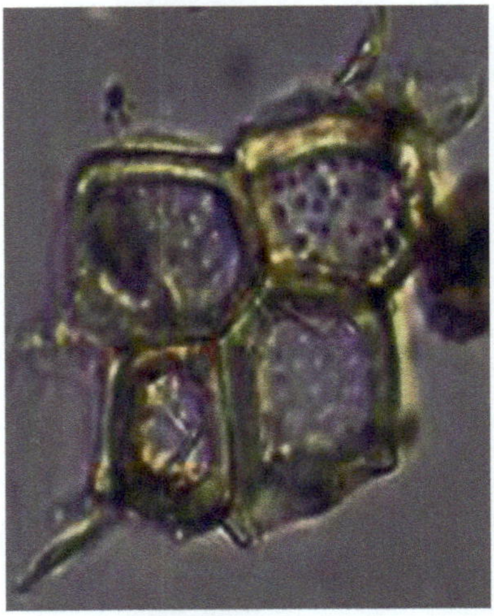

FIG. 5.12.6 Pitted parenchyma from *Crataeva nurvula* bark powder.

References

[1] Government of India, Department of AYUSH, Ayurvedic Pharmacopoeia of India. Part I, vol. I, Ministry of Health and welfare, Ghaziabad, 1999–2017, p. 159.

[2] A.K. Gupta, N. Tandon, M. Sharma, Quality Standards of Indian Medicinal Plants, vol. 10, Medicinal plant unit, ICMR, New Delhi, India, 2008, p. 107.

Chapter 5.13

Litsea chinensis

Litsea chinensis Lam. Lauraceae

Common vernacular names: Bengali – Kukurchite; Gujrati – Meda Lakdee; Hindi – Maida Lakdee; Marathi – Meda Lakdee; Punjabi – Medasaka; Tamil – Medalakavi; Telugu – Meda [1] (Fig. 5.13.1).

FIG. 5.13.1 *Litsea chinensis* bark.

Macroscopic description: Shape – Channeled cut pieces; **Outer bark** – Rough, corky, brown; **Inner bark** – Smooth, longitudinally striated, Dark brown to black; **Fracture** – Short and uneven; **Odour** – Characteristic.

Major chemical constituents: Alkaloids (Laurotetaline, actinodaphine, boldine, norboldine, sebiferine and litseferine) [1].

Traditional uses: Wound healing and stomach problems [1].

Powder microscopic description: Light brown powder showing abundant fibrous sclereids isolated or usually in groups, thick walled with distinct pits and striations, wide to narrow lumen, few with peg like extension; Oval

Powdered Crude Drug Microscopy of Leaves and Barks. https://doi.org/10.1016/B978-0-12-818092-1.00017-8

rectangular to triangular in shape-various shape and size; abundant stone cells, oval to squarish in shape with wide or narrow lumen, with distinct pits, striated walls and pitted lumen; prismatic crystals of calcium oxalate and simple, small, spherical starch grains; fragments of cork cells - polygonal, pentagonal to hexagonal in shape and fragments of thick walled fibres with pointed apex. Overnight soaked powder become very sticky showing that bark is highly mucilaginous. Moreover, it is difficult to handle sample with a brush as it is highly viscous and sticky. Features are clearly visible only after potassium chlorate treatment. The general as well as specific features of *L. chinesis* bark powder are illustrated in Table 5.13.1 and Figs. 5.13.2–5.13.7 respectively.

TABLE 5.13.1 Specific as well as general characters of bark powder of *L. chinesis.*

	Specific features			General features
	Stone cells	**Sclereids**	**Fibres**	**Calcium oxalate crystals**
Occurrence	Isolated	Isolated	Numerous	Prismatic
	Group	Group	Isolated	
	Abundant	Abundant		**Cork**
Walls	Distinct pits	Thick walled	Thick walled	Hexagonal
	Striated walls	Radiating pits	Thin walled	Pentagonal
	Thin walled	Peg like extension	Smoothly striated	Beaded walls
	Thick walled			Polygonal
Shape	Oval	Oval	Very long	Irregular
	Squarish	Rectangular	Pointed apex	
	Rectangular	Triangular	Irregular margin	**Starch grains**
	Triangular	Fibre like		Simple
Lumen	Wide	Wide	Wide	Compound
	Narrow	Narrow	Narrow	Oval
	Pitted	Pitted		Round
	Slit like			Small

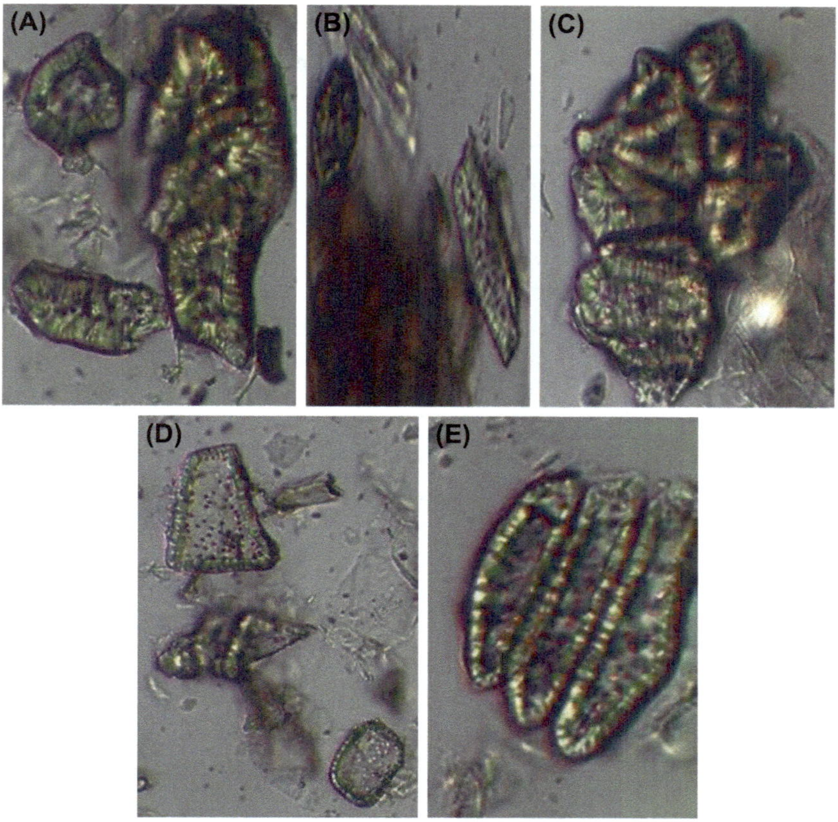

FIG. 5.13.2 (A–E) – Stone cells of various shapes and sizes from *Litsea chinensis* bark powder.

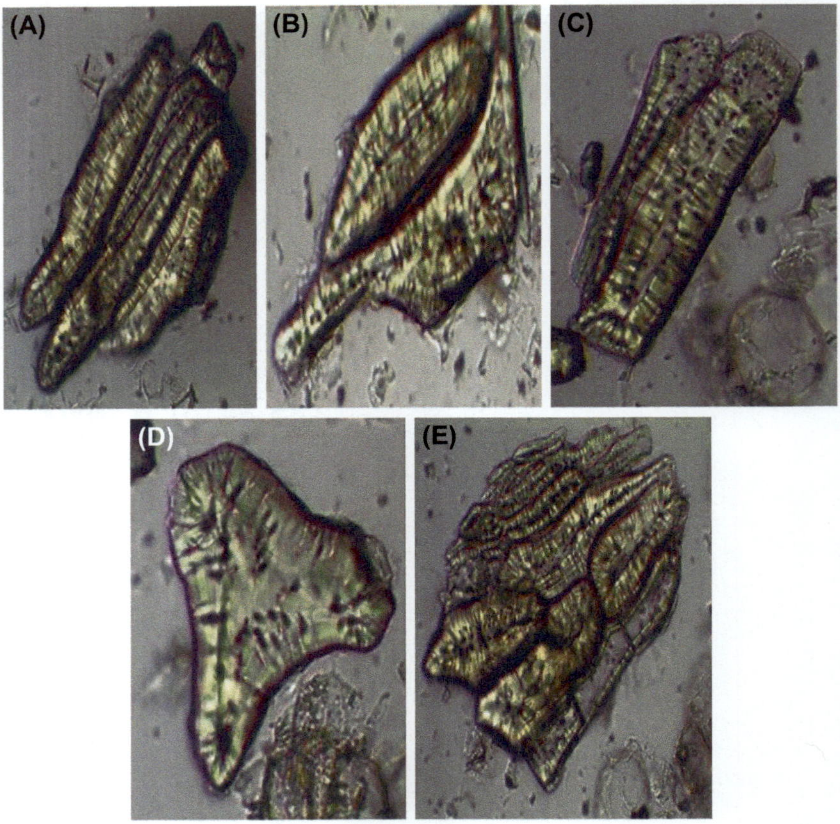

FIG. 5.13.3 (A–E) – Sclereids of various shapes and sizes from *Litsea chinensis* bark powder.

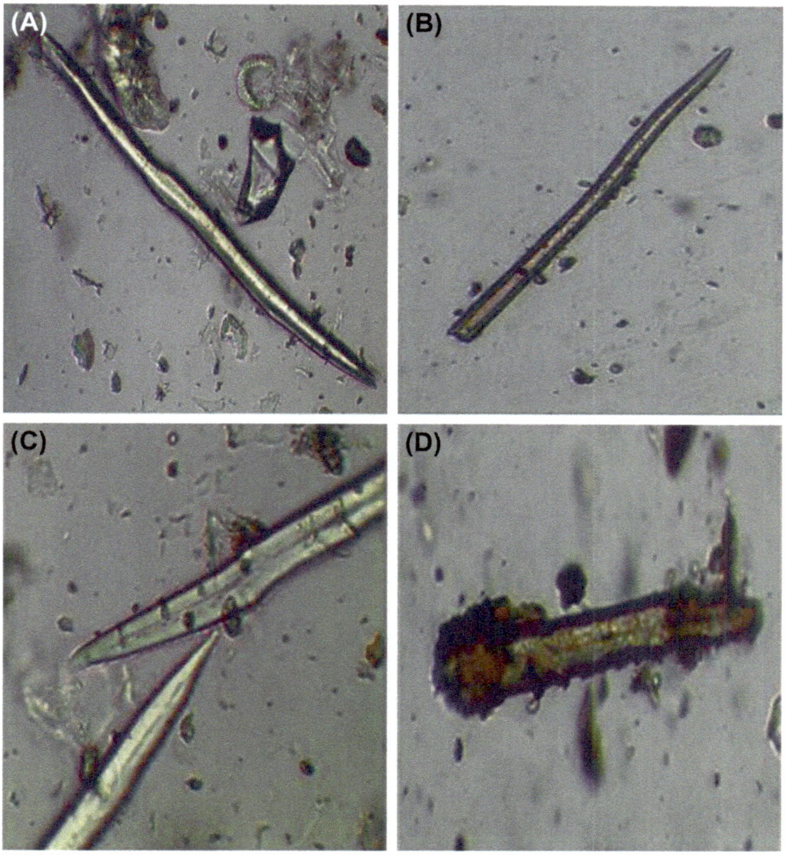

FIG. 5.13.4 (A–D) – Fibres of various shapes and sizes from *Litsea chinensis* bark powder.

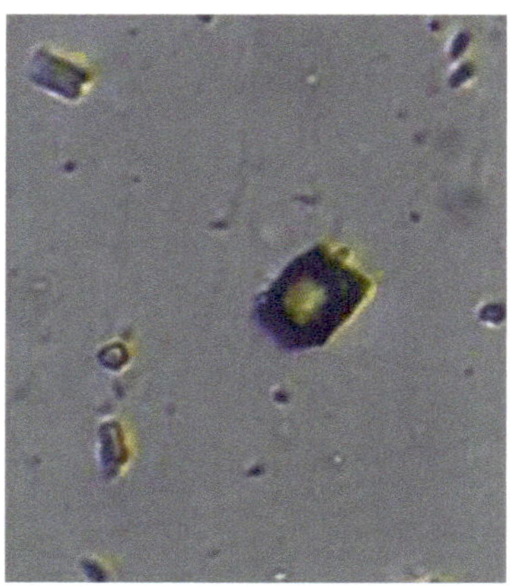

FIG. 5.13.5 Prismatic crystals from *Litsea chinensis* bark powder.

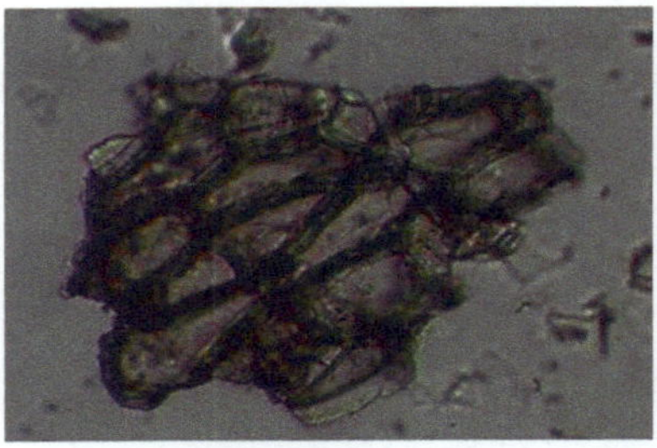

FIG. 5.13.6 Cork cells from *Litsea chinensis* bark powder.

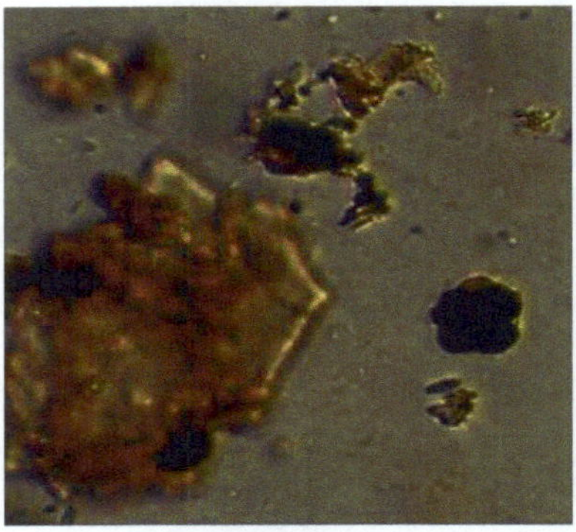

FIG. 5.13.7 Starch grains of from *Litsea chinensis* bark powder.

Reference

[1] Government of India, Department of AYUSH, Ayurvedic Pharmacopoeia of India. Part I, vol. V, Ministry of Health and welfare, Ghaziabad, 1999–2017, p. 124.

Chapter 5.14

Mangifera indica

Mangifera indica L. Anacardiaceae

Common vernacular names: Assamese – Aam; Bengali – Ama, Am; English – Mango; Gujrati – Ambo; Hindi – Ama; Kannada – Mavu; Malyalam – Mavu; Marathi – Amba; Oriya – Am, Amba; Punjabi – Amb; Tamil – Mamaram; Telugu – Amaramu [1] (Fig. 5.14.1).

FIG. 5.14.1 *Mangifera indica* bark.

Macroscopic description: Shape – Pieces of variable size and thickness; **Outer bark** – Dark brown, rough due to longitudinal cracks and fissures; **Inner bark** – Yellowish white to reddish; **Fracture** – Fibrous; **Odour** – Pleasant.

Major chemical constituents: Tannins - Protocatechuic Acid, Catechin, Mangiferin, Alanine, Glycine, α -Aminobutyric acid, Kinic and Shikimic Acids [1].

Traditional uses: Dentrifice, antiseptic and stomachic [1].

Powdered Crude Drug Microscopy of Leaves and Barks. https://doi.org/10.1016/B978-0-12-818092-1.00018-X

Powder microscopic description: Brown coloured powder showing very abundant prismatic crystals of calcium oxalate of various shape and sizes scattered as such; abundant, very long fibres, isolated or in groups exhibiting fine reticulations, narrow lumen often getting abruptly widened at places, some of them are with pointed apex; fragments of cork cells-polygonal to pentagonal and stone cells, oval to polygonal, thick walled, narrow lumen and striated walls. The general as well as specific features of *M. indica* bark powder are illustrated in Table 5.14.1 and Fig. 5.14.2–5.14.5 respectively.

TABLE 5.14.1 Specific as well as general characters of bark powder of *M. indica*.

	Specific features		General features
	Stone cells	**Fibres**	**Calcium oxalate crystals**
Occurrence	Few	Isolated	Prismatic
	Group	Group	
	Isolated	Numerous	**Cork**
Walls	Thick walled	Smoothly striated	Polygonal
	Thin walled	Thick walled	Pentagonal
	Striations	Bifurcating ends	
	Radiating pits	Septate	
Shape	Oval	Very long	
	Polygonal	Pointed apex	
	Spherical	Irregular margin	
	Squarish	Bent	
Lumen	Narrow	Narrow	
	Pitted		
	Wide		

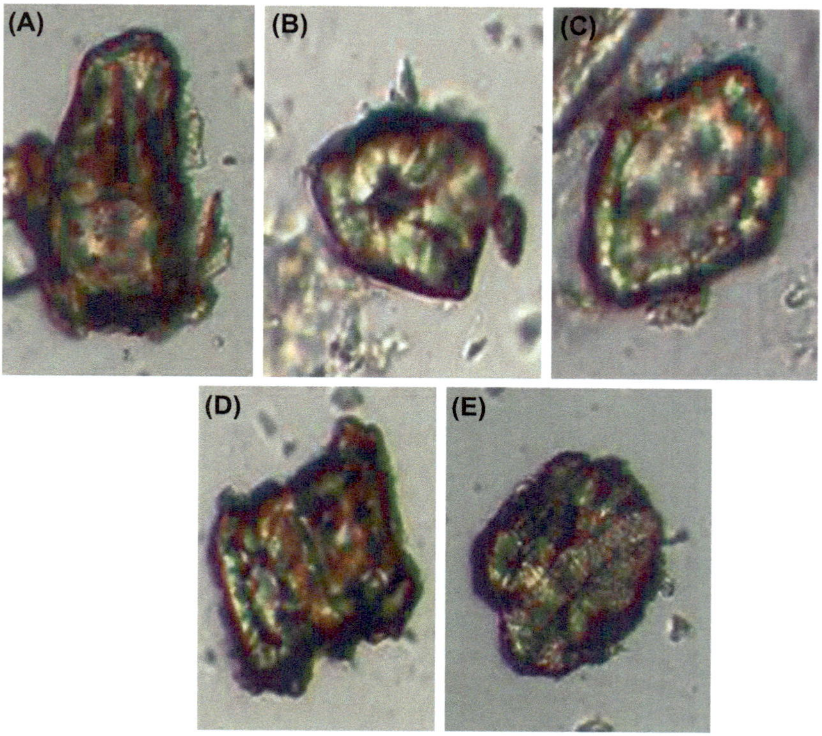

FIG. 5.14.2 (A–E) – Stone cells of various shapes and sizes from *Mangifera indica* bark powder.

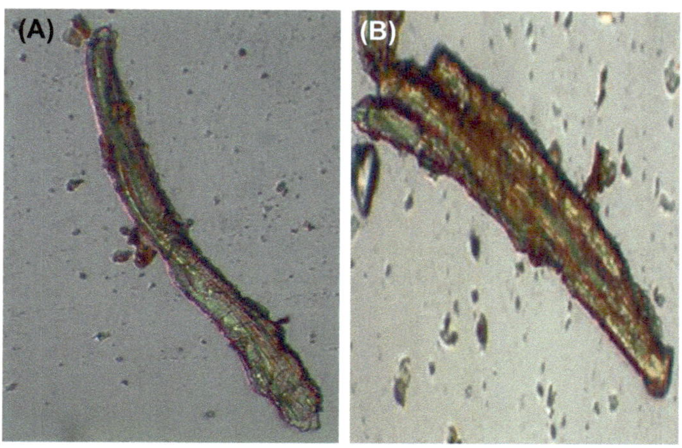

FIG. 5.14.3 (A–B) – Fibres of various shapes and sizes from *Mangifera indica* bark powder.

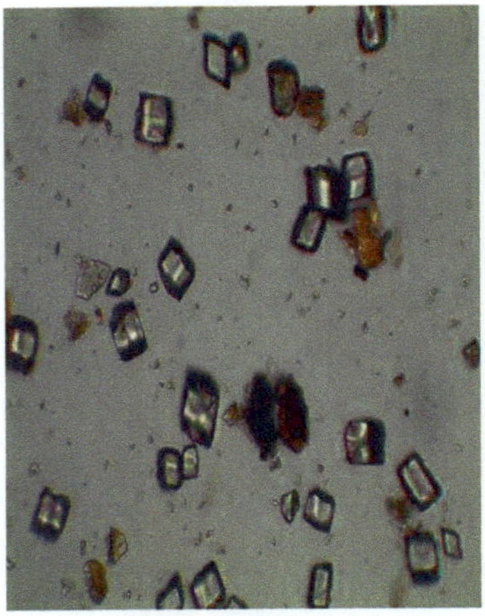

FIG. 5.14.4 Prismatic crystals from *Mangifera indica* bark powder.

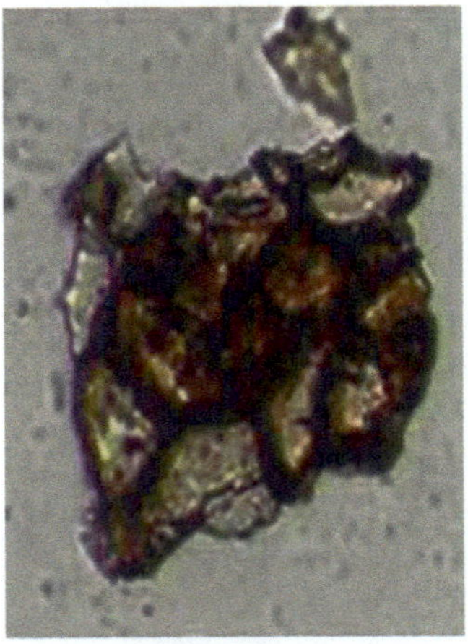

FIG. 5.14.5 Cork cells from *Mangifera indica* bark powder.

Reference

[1] Government of India, Department of AYUSH, Ayurvedic Pharmacopoeia of India. Part I, vol. III, Ministry of Health and welfare, Ghaziabad, 1999–2017, p. 10.

Chapter 5.15

Moringa oleifera

Moringa oleifera Lam. Moringaceae

Common vernacular names: Bengali – Sajina, Sajne; English – Horse Radish Tree, Drum-stick Tree; Gujrati – Saragave; Hindi – Sahijana; Kannada – Nugge, Nuggemara, Nuggekoyimara; Malayalam – Muringya, Murinna; Marathi – Shewga; Oriya – Munga, Munika, Sajana; Punjabi – Sohajana; Tamil – Murungai; Telugu – Munaga chettu, Mulaya Chetta [1] (Fig. 5.15.1).

FIG. 5.15.1 *Moringa oleifera* bark.

Macroscopic description: Shape – Channeled pieces; **Outer bark** – Greyish brown, longitudinally fissured; **Inner bark** – Buff in colour, faintly longitudinally striated; **Fracture** – Outer short, inner splintery; **Odour** – Characteristic.

Powdered Crude Drug Microscopy of Leaves and Barks. https://doi.org/10.1016/B978-0-12-818092-1.00019-1

Major chemical constituents: Moringine, moringinine, trans-4-hydroxymellein, vanillin [2].

Traditional uses: Colic, dermatitis and worm infestation [2].

Powder microscopic description: Light greyish brown colour powder showing fragments of thick walled, polygonal to hexagonal cork cells; numerous fragments of fibres, isolated or in group, varying in size and shape; abundant, isolated or groups of stone cells and few sclereids varying in size, shape and thickness; fragments of radially-longitudinally cut medullary rays crossing the fibre; plenty of prismatic and few cluster crystal of calcium oxalate and simple, spherical to oval starch grains scattered as such [3]. The general as well as specific features of *M. oleifera* bark powder are illustrated in Table 5.15.1 and Figs. 5.15.2–5.15.6 respectively.

TABLE 5.15.1 Specific as well as general characters of bark powder of *M. oleifera.*

	Specific features			General features
	Stone cells	**Sclereids**	**Fibres**	**Calcium oxalate crystals**
Occurrence	Abundant	Few	Numerous	Prismatic
	Isolated	Isolated	Isolated	Cluster
	Group		Group	
Walls	Striations	Thick walled	Thick walled	
	Thick walled	Striations	Smoothly striated	**Cork**
			Faint transverse septa	Polygonal
Shape	Oval	Spherical	Pointed apex	Pentagonal
	Angular	Elongated	Irregular margin	Hexagonal
			Very long	Thick walled
			Bent	
			Rectangular	
Lumen	Narrow	Narrow	Narrow	
		Pitted		

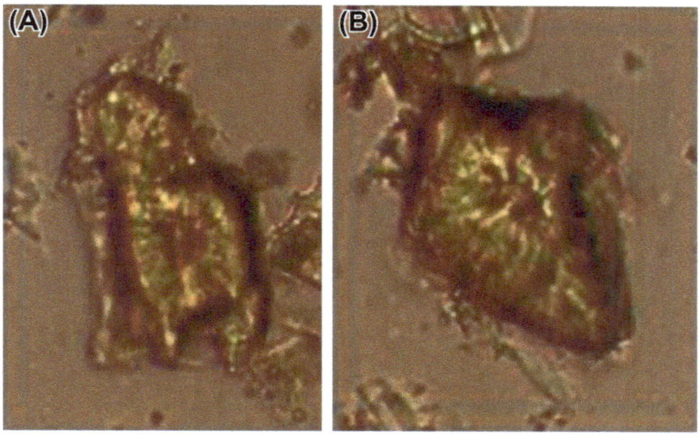

FIG. 5.15 2 (A–B) – Stone cells of various shapes and sizes from *Moringa oleifera* bark powder.

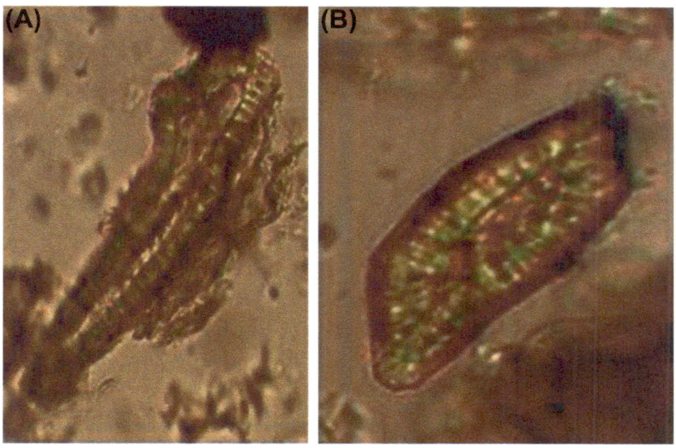

FIG. 5.15 3 (A–B) – Sclereids of various shapes and sizes from *Moringa oleifera* bark powder.

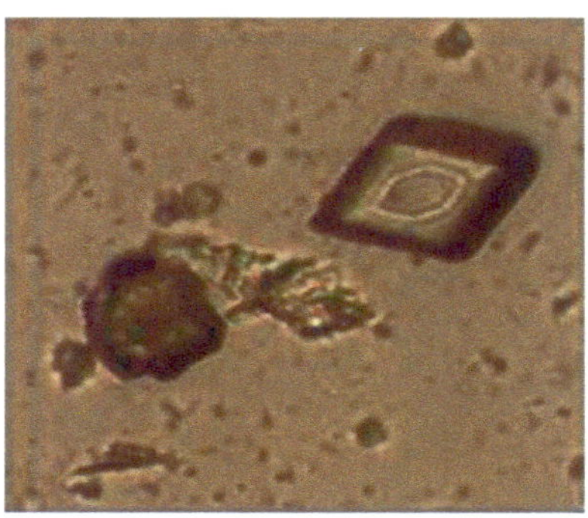

FIG. 5.15.4 Prismatic crystals from *Moringa oleifera* bark powder.

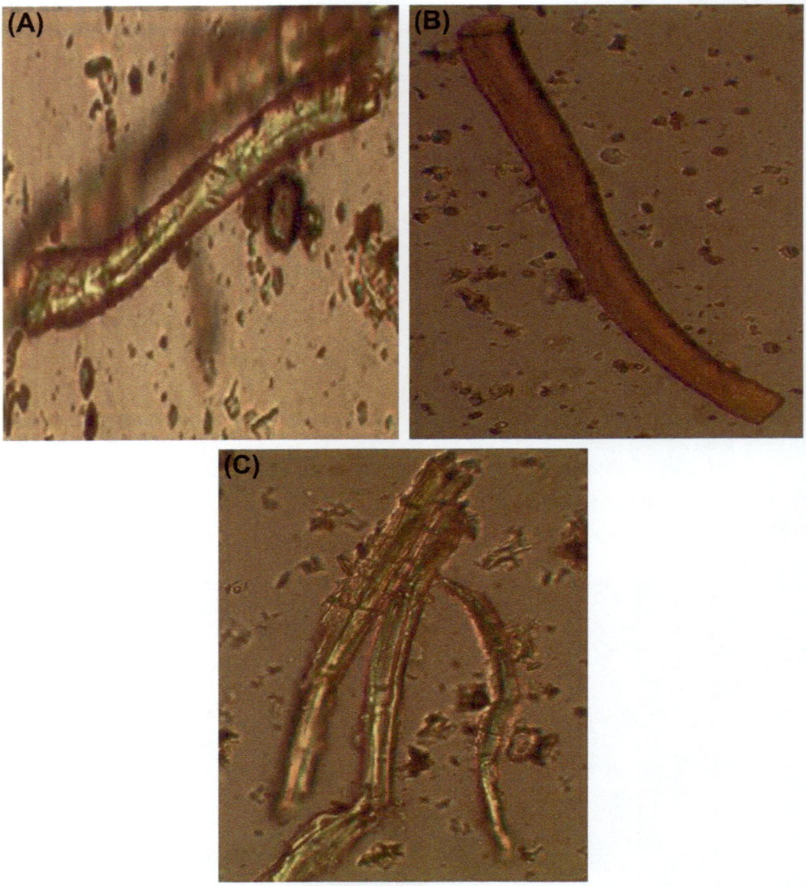

FIG. 5.15.5 (A–C) – Fibres of various shapes and sizes from *Moringa oleifera* bark powder.

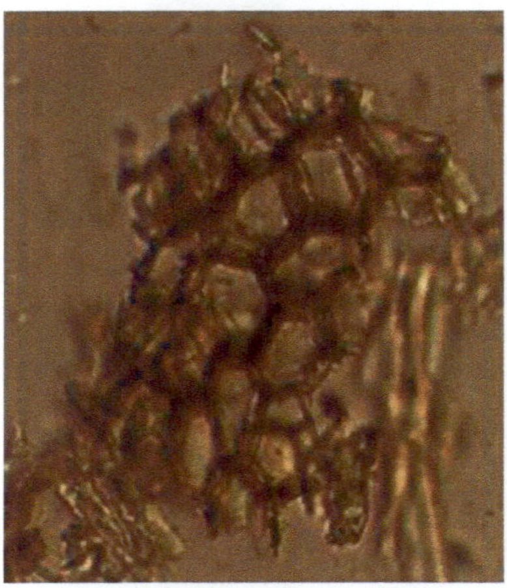

FIG. 5.15.6 Cork cells from *Moringa oleifera* bark powder.

References

[1] Government of India, Department of AYUSH, Ayurvedic Pharmacopoeia of India. Part I, vol. IV, Ministry of Health and welfare, Ghaziabad, 1999–2017, p. 130.

[2] A.K. Gupta, N. Tandon, M. Sharma, Quality Standards of Indian Medicinal Plants, vol. 8, Medicinal plant unit, ICMR, New Delhi, India, 2008, p. 203.

[3] M.G. Chauhan, A.P.G. Pillai, Microscopic Profile of Powdered Drugs Used in Indian System of Medicine, Bark Drugs, vol. 1, Institute of Ayurvedic Medicinal Plant Sciences, Jamnagar, Gujarat, India, 2005, p. 92.

Chapter 5.16

Pongamia pinnata

Pongamia pinnata (L.) Pierre Fabaceae

Common vernacular names: Assamese – Korach; Bengali – Dahara karanja, Karanja, Natakaranja; Gujrati – Kanaji; Hindi – Karanj; Kannada – Honge; Malayalam – Pungu, Ungu; Marathi – Karanja; Oriya – Karanja; Punjabi – Karanj; Tamil – Pungai; Telugu – Ganuga, Kanuga [1] (Fig. 5.16.1).

FIG. 5.16.1 *Moringa oleifera* bark.

Macroscopic description: Shape – Channeled, recurved and thick pieces; **Outer bark** – Ash grey to greyish brown; **Inner bark** – Yellowish white to cream colour; **Fracture** – Short and fibrous; **Odour** – Unpleasant.

Major chemical constituents: Flavones and Furanoflavones like Karanjin, Pongapin, Demethoxykanugin, Kanugin, Pinnatin, Tetra-o-Methylfisetin, Gamatin, 5-Methoxyfurano (2″, 3″ 7:8), flavone and 5-Methoxy-3′4′

Powdered Crude Drug Microscopy of Leaves and Barks. https://doi.org/10.1016/B978-0-12-818092-1.00020-8

Methylene dioxyfurano (2″, 3″, 7:8) flavone & two new Furano compounds Glabra-I and Glabra-II [1].

Traditional uses: Piles, Diabetes, antimicrobial [1].

Powder microscopic description: Yellowish cream powder showing abundant large group of stone cells of various sizes, shapes and thickness squarish, oblong, rectangular, oval spherical in shape with wide lumen and branching pits; abundant prismatic crystals of calcium oxalate and few sclereids rectangular in shape [2]. The general as well as specific features of *P. pinnata* bark powder are illustrated in Table 5.16.1 and Figs. 5.16.2–5.16.4 respectively.

TABLE 5.16.1 Specific as well as general characters of bark powder of *P. pinnata.*

	Specific features		General features
	Stone cells	**Sclereids**	**Calcium oxalate crystals**
Occurrence	Isolated	Few	Prismatic
	Group	Isolated	
	Abundant	Group	
Walls	Thin walled	Striations	
	Branching pits	Thick walled	
Shape	Squarish	Rectangular	
	Oblong	Oval	
	Oval	Spherical	
	Spherical		
Lumen	Wide	Narrow	

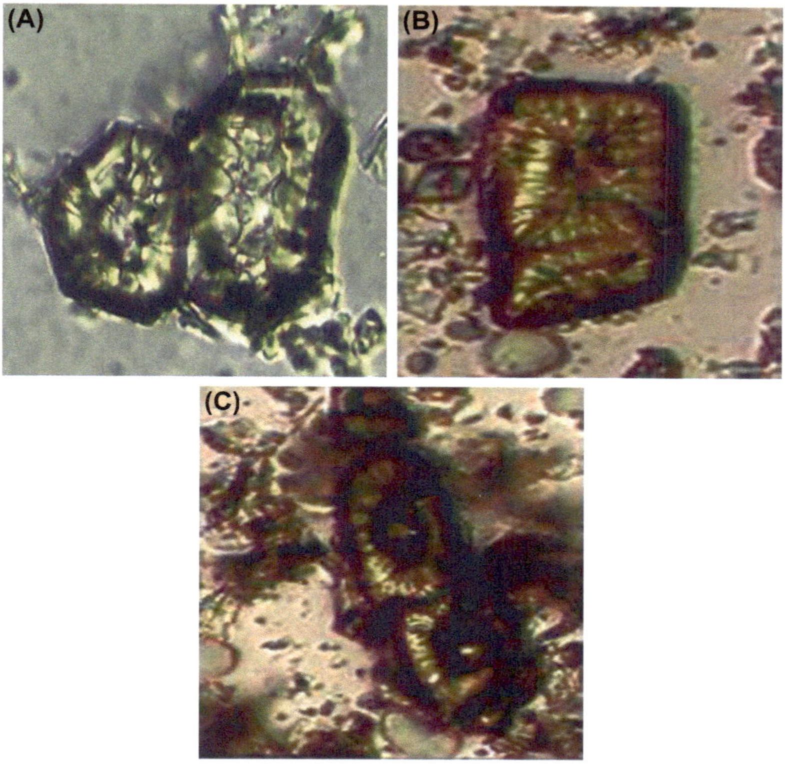

FIG. 5.16.2 (A–C) – Stone cells of various shapes and sizes from *Pongamia pinnata* bark powder.

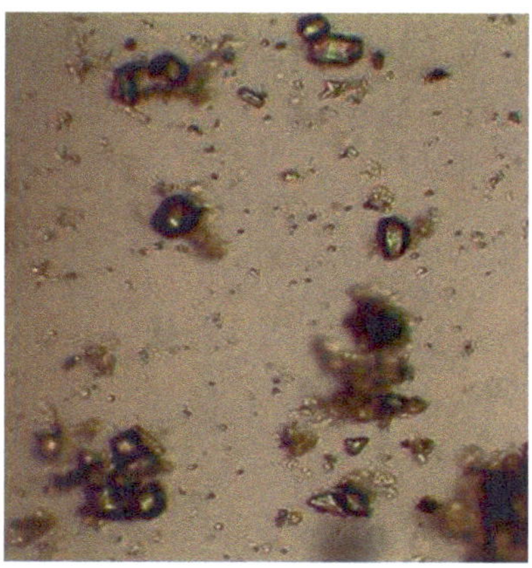

FIG. 5.16.3 Prismatic crystals from *Pongamia pinnata* bark powder.

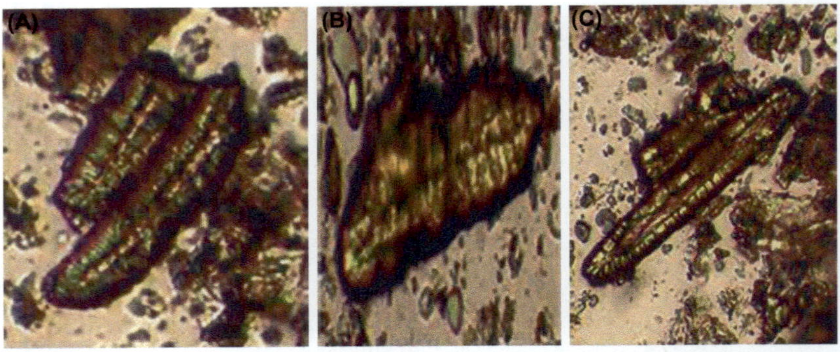

FIG. 5.16.4 (A–C) – Sclereids of various shapes and sizes from *Pongamia pinnata* bark powder.

References

[1] Government of India, Department of AYUSH, Ayurvedic Pharmacopoeia of India. Part I, vol. II, Ministry of Health and welfare, Ghaziabad, 1999–2017, p. 85.

[2] M.G. Chauhan, A.P.G. Pillai, Microscopic Profile of Powdered Drugs Used in Indian System of Medicine, Bark Drugs, vol. 1, Institute of Ayurvedic Medicinal Plant Sciences, Jamnagar, Gujarat, India, 2005, p. 54.

Chapter 5.17

Spondias pinnata

Spondias pinnata (L. f.) Kurz Anacardiaceae

Common vernacular names: Bengali – Amada, Amra; English – Indian Hog Plum, Wild Mango; Gujarati – Ambeda, Ambado, Ranamba, Jangali Ambo, Ranambo; Hindi – Ambada, Amra, Jangli Aam; Kannada – Ambate, Amvara; Malayalam – Mampusli, Ambalam, Ambazham, Mampuiti, Ampozham, Njettikuzhiyan mavu; Marathi – Ambado; Punjabi – Amada; Tamil – Mambulichi Amputtai, Ambadam; Telugu – Amratakamu, Anbalamu, Adavimamidi [1] (Fig. 5.17.1).

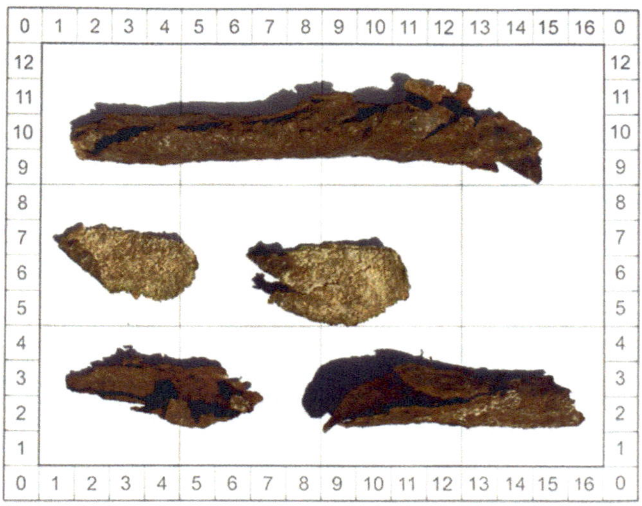

FIG. 5.17.1 *Spondias pinnata* bark.

Powdered Crude Drug Microscopy of Leaves and Barks. https://doi.org/10.1016/B978-0-12-818092-1.00021-X

Macroscopic description: Shape – Thin, curved and long cut pieces; **Outer bark** – Smooth, grey having lenticels; **Inner bark** – Reddish yellow; **Fracture** – Laminated; **Odour** – Characteristic.

Major chemical constituents: Tannins and starch [1].

Traditional uses: Refrigerant, Tonic, Antiseptic, Astringent [1].

Powder microscopic description: Light reddish yellow colour powder showing abundant stone cells of varying shapes and sizes; few prismatic crystals; cork cells – thick walled, polygonal; few sclereids and simple fibres. Microscopical features are not very clear in overnight soaked powder, but potassium chlorate treatment enhanced them. The general as well as specific features of *S. pinnata* bark powder are illustrated in Table 5.17.1 and Figs. 5.17.2–5.17.6 respectively.

TABLE 5.17.1 Specific as well as general characters of bark powder of *S. pinnata*.

	Specific features			General features
	Stone cells	**Sclereids**	**Fibres**	**Calcium oxalate crystals**
Occurrence	Abundant	Few	Few	Prismatic
	Isolated	Isolated	Isolated	
	Group			**Cork**
Walls	Striations	Striations	Thick walled	Thick walled
	Thick walled	Thick walled		Polygonal
	Thin walled			Hexagonal
Shape	Oval	Spherical	Very long	
	Circular	Irregular	Rectangular	
	Spherical			
Lumen	Narrow	Narrow	Narrow	
	Slit like			

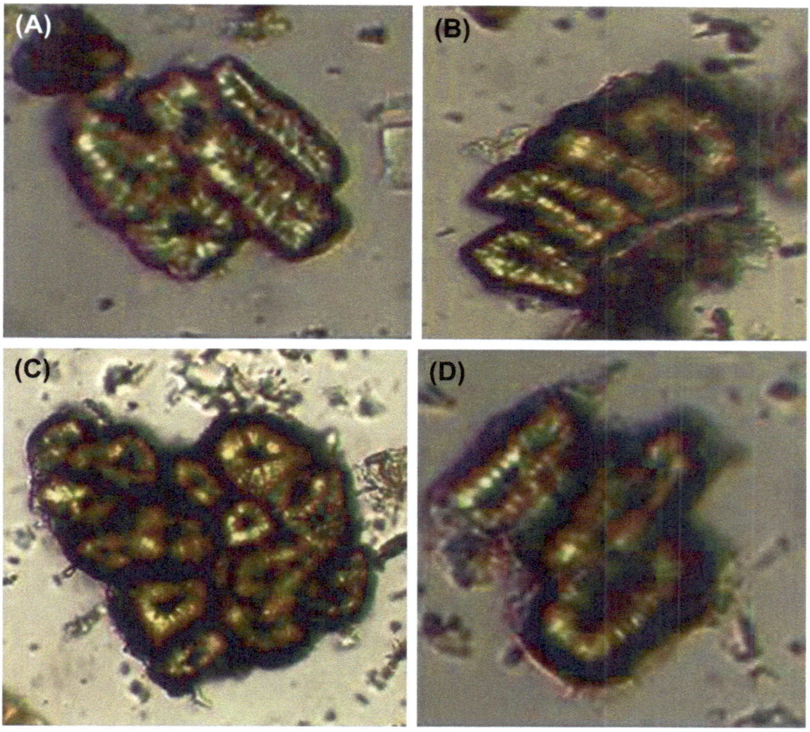

FIG. 5.17.2 (A–D) – Stone cells of various shapes and sizes from *Spondias pinnata* bark powder.

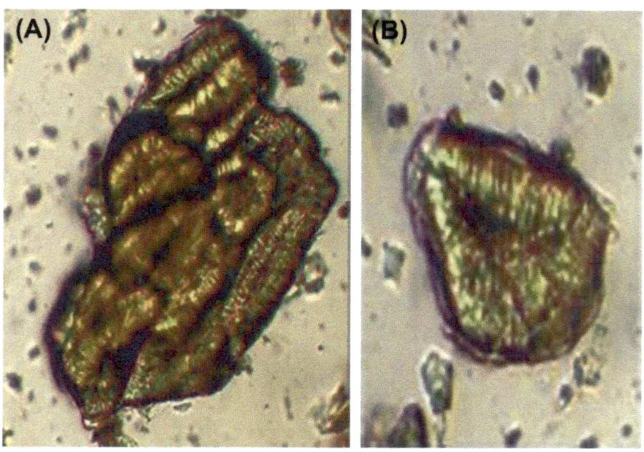

FIG. 5.17.3 (A–B) – Sclereids of various shapes and sizes from *Spondias pinnata* bark powder.

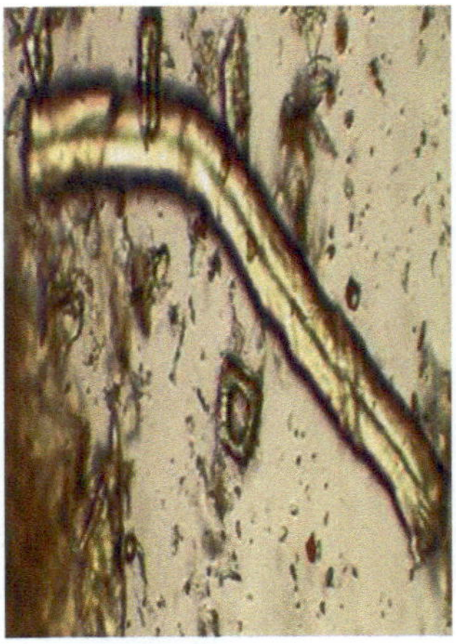

FIG. 5.17.4 Fibre from *Spondias pinnata* bark powder.

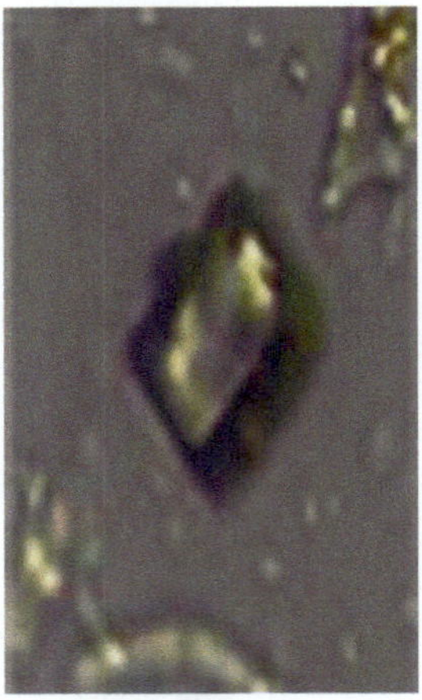

FIG. 5.17.5 Prismatic crystal from *Spondias pinnata* bark powder.

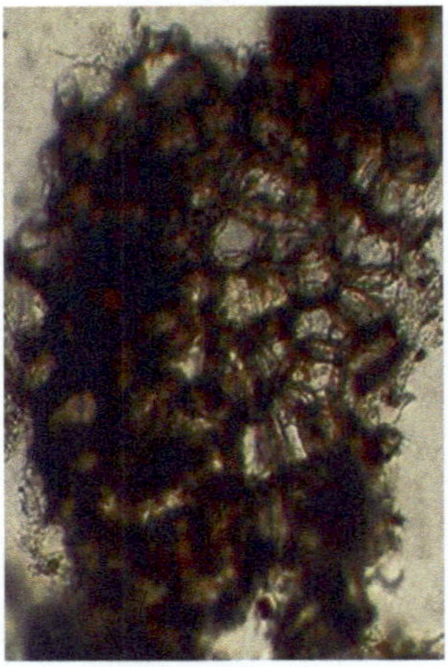

FIG. 5.17.6 Cork cells from *Spondias pinnata* bark powder.

Reference

[1] Government of India, Department of AYUSH, Ayurvedic Pharmacopoeia of India. Part I, vol. II, Ministry of Health and Welfare, Ghaziabad, 1999–2017, p. 06.

Chapter 5.18

Stereospermum chelonoides

Stereospermum chelonoides (L. f.) DC. Bignoniaceae

Common vernacular names: Bengali – Paarul; English – Trumpet Flower Tree, Yellow Snake Tree; Gujarati – Paadal; Hindi – Paraal, Paatar, Paadree, Paadhal; Kannada – Rude, Kalludi, Kaala-adri; Malayalam – Pup-paatiri, Paatiri; Marathi – Paadal; Oriya – Patudi; Punjabi – Paadal; Tamil – Paadiri, Pumpaadiri, Paadari; Telugu – Kokkosa, Kaligottu [1] (Fig. 5.18.1).

FIG. 5.18.1 *Stereospermum chelonoides* bark.

Macroscopic description: Shape – Slightly recurved and channelled; **Outer bark** – Dull brown, rough due to presence of lenticels and fissures; **Inner bark** – Almost smooth, light brown and finely striated; **Fracture** – Tough and short with inner lamellae peeling off; **Odour** – Not characteristic.

Major chemical constituents: Triacontanol [2].

Traditional uses: Dyspnoea and anti-inflammatory [2].

Powdered Crude Drug Microscopy of Leaves and Barks. https://doi.org/10.1016/B978-0-12-818092-1.00022-1

Powder microscopic description: Brown colour powder showing prismatic, rosette and cluster crystals scattered as such; abundant fragments of cork cells – polygonal to pentagonal to hexagonal in shape; abundant stone cells of various shape and sizes; simple and compound starch grains observed on Iodine solution treatment and fragments of medullary rays. The general as well as specific features of *S. chelonoides* bark powder are illustrated in Table 5.18.1 and Figs. 5.18.2–5.18.7 respectively.

TABLE 5.18.1 Specific as well as general characters of bark powder of *S. chelonoides*.

	Specific features	General features
	Stone cells	**Calcium oxalate crystals**
Occurrence	Abundant	Prismatic
	Isolated	Rosette
	Group	Cluster
Walls	Thin walled	**Cork**
	Thick walled	Thick walled
	Striations	Polygonal
	Radiating pits	Pentagonal
	Distinct pits	Hexagonal
Shape	Oval	**Starch grains**
	Triangular	Simple
	Spherical	Compound
	Circular	Oval
	Oblong	**Medullary rays**
	Pentagonal	Biseriate
Lumen	Wide	Triseriate
	Narrow	
	Pitted	

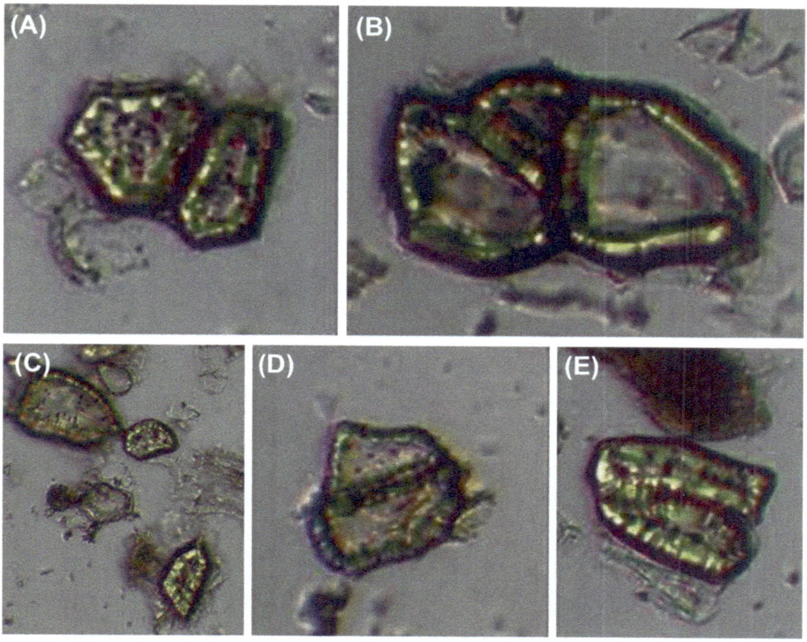

FIG. 5.18.2 (A–E) – Stone cells of various shapes and sizes from *Stereospermum chelonoides* bark powder.

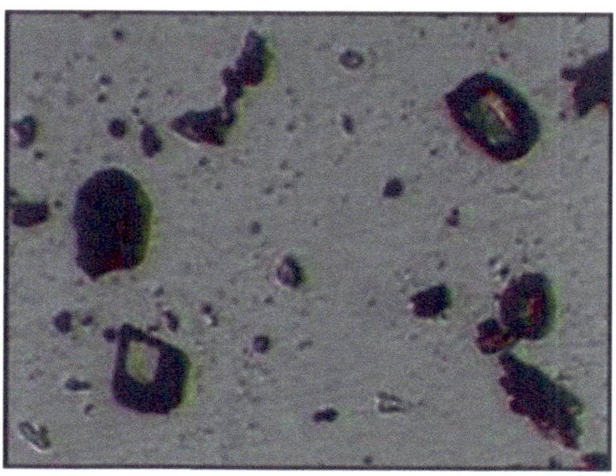

FIG. 5.18.3 Prismatic crystals from *Stereospermum chelonoides* bark powder.

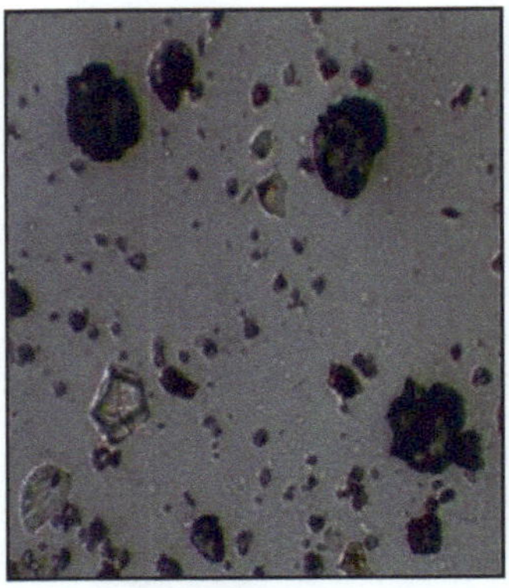

FIG. 5.18.4 Rosette crystals from *Stereospermum chelonoides* bark powder.

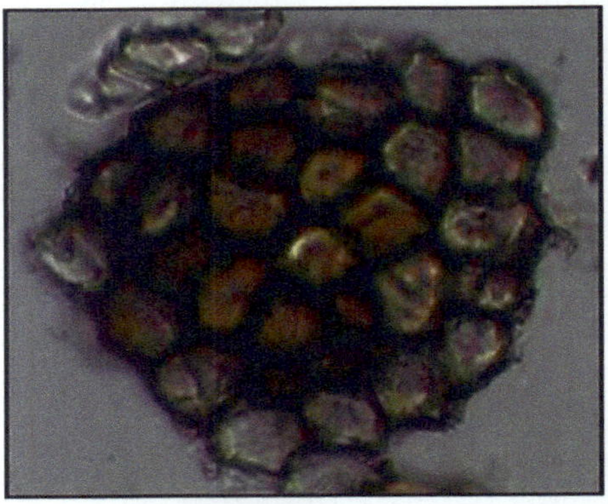

FIG. 5.18.5 Cork cells from *Stereospermum chelonoides* bark powder.

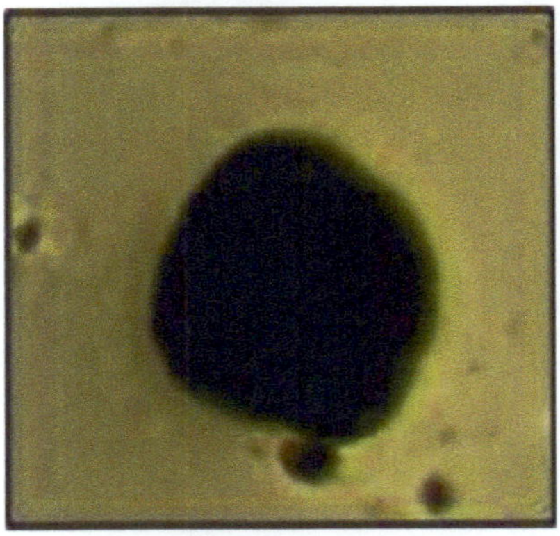

FIG. 5.18.6 Starch grains from *Stereospermum chelonoides* bark powder.

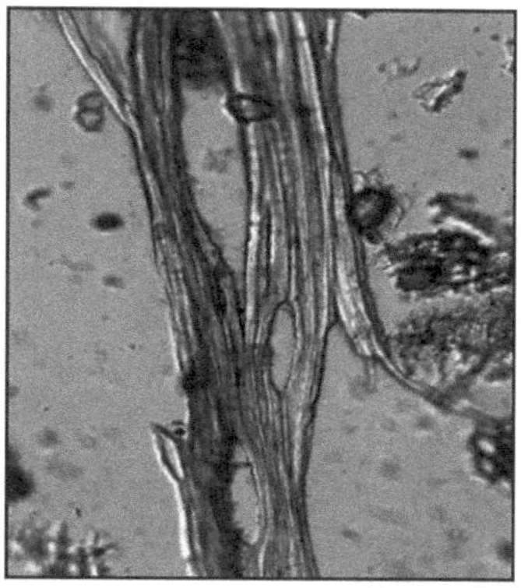

FIG. 5.18.7 Medullary rays from *Stereospermum chelonoides* bark powder.

References

[1] Government of India, Department of AYUSH, Ayurvedic Pharmacopoeia of India. Part I, vol. IV, Ministry of Health and Welfare, Ghaziabad, 1999–2017, p. 100.

[2] A.K. Gupta, N. Tandon, M. Sharma, Quality Standards of Indian Medicinal Plants, vol. 7, Medicinal Plant Unit, ICMR, New Delhi, India, 2008, p. 287.

Chapter 5.19

Syzygium cumini

Syzygium cumini (L.) Skeels Myrtaceae

Common vernacular names: Assamese – Jam; Bengali – Jaam; Gujarati – Jambu, Jambuda; Hindi – Jomuna, Raja Jambu; Kannada – Merale, Jamneralae, Jambu, Neralamara; Malayalam – Njaval, Naval; Marathi – Jambhool; Oriya – Jamukoli, Jamu, Jam; Punjabi – Jammu; Tamil – Naaval, Navval Sambu, Mahamaram, Nagal; Telugu – Nesedu [1] (Fig. 5.19.1).

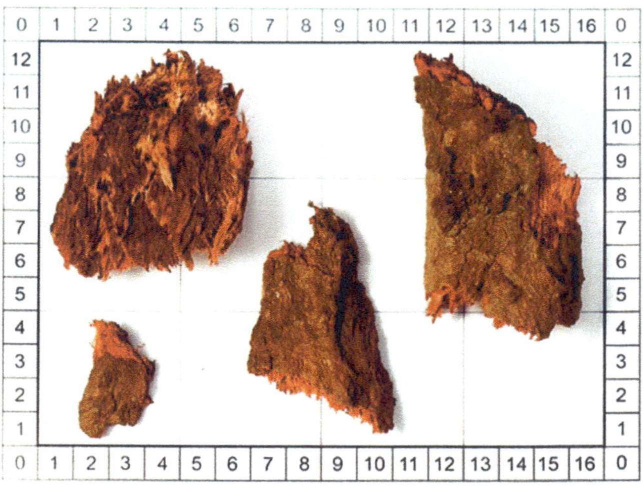

FIG. 5.19.1 *Syzygium cumini* bark.

Macroscopic description: Shape – Channeled, curved or flat pieces; **Outer bark** – Light greyish brown, more or less rough and rugged due to exfoliation and vertical cracks; **Inner bark** – Reddish brown, fibrous and rough; **Fracture** – Short and splintery; **Odour** – Characteristic.

Major chemical constituents: Betulinic acid, Friedelin, epi-friedelanol, β-sitosterol, Eugenin [1].

Traditional uses: Diabetes, Dysentery, Miscarriage [1].

Powdered Crude Drug Microscopy of Leaves and Barks. https://doi.org/10.1016/B978-0-12-818092-1.00023-3

Powder microscopic description: Light brown colour powder showing abundant sclereids of various sizes and shapes, mostly isolated, fibrous or branched with thick pitted striated wall and with narrow wide lumen; stone cells, small in comparison to sclereids, isolated or in groups, varying in size, circular, oval to squarish in shape, occasionally three sided thickened with pitted or striated wall and wide or narrow lumen; fragments of thick walled fibres with bifurcating ends; abundant rosette and cluster crystals of calcium oxalate scattered as such throughout; starch grains, simple and compound, oval in shape with eccentric hilum and fragments of tangentially and radially longitudinally cut medullary rays [2]. The general as well as specific features of *S. cumini* bark powder are illustrated in Table 5.19.1 and Figs. 5.19.2–5.19.6 respectively.

TABLE 5.19.1 Specific as well as general characters of bark powder of *S. cumini*.

	Specific features			General features
	Sclereids	**Stone cells**	**Fibres**	**Calcium oxalate crystals**
Occurrence	Isolated	Few	Isolated	Rosette
	Group	Group	Group	
	Abundant		Numerous	**Starch grains**
Walls	Striations	Striations	Thick walled	Few
	Thick walled	Radiating pits	Septate	Simple
	Projections	Thin walled	Bifurcating ends	Compound
	Thin walled		Striations	Oval
Shape	Oval	Oval	Rectangular	Eccentric hilum
	Rectangular	Circular	Bent	
	Blunt end			
Lumen	Narrow	Pitted	Narrow	
	Pitted	Wide	Slit like	
	Wide			

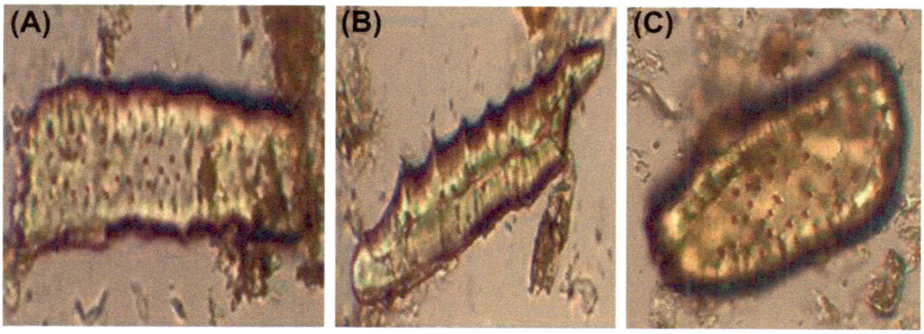

FIG. 5.19.2 (A–C) – Sclereids of various shapes and sizes from *Syzygium cumini* bark powder.

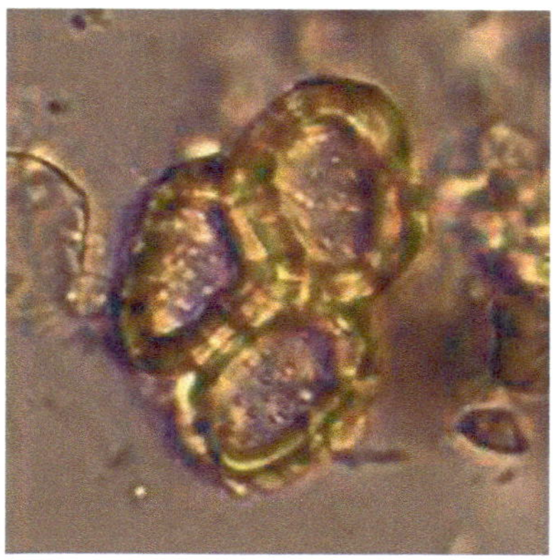

FIG. 5.19.3 Stone cells from *Syzygium cumini* bark powder.

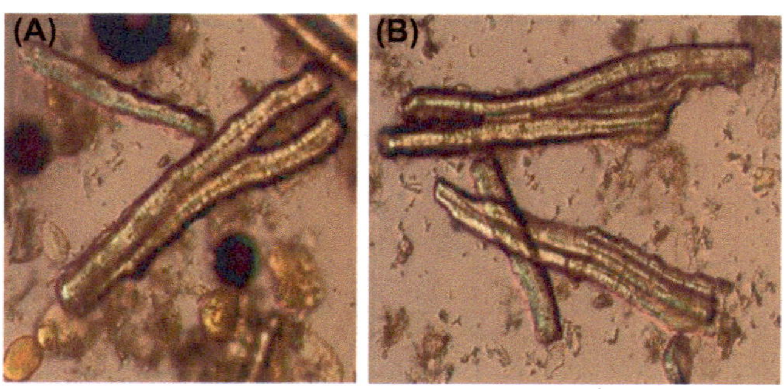

FIG. 5.19.4 (A–B) – Fibres of various shapes and sizes from *Syzygium cumini* bark powder.

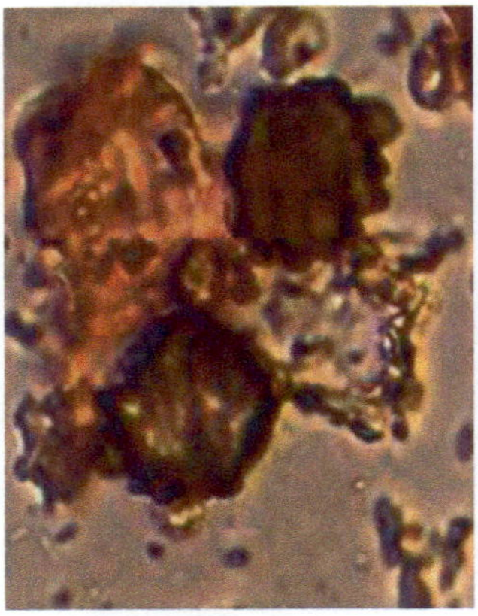

FIG. 5.19.5 Rosette crystals from *Syzygium cumini* bark powder.

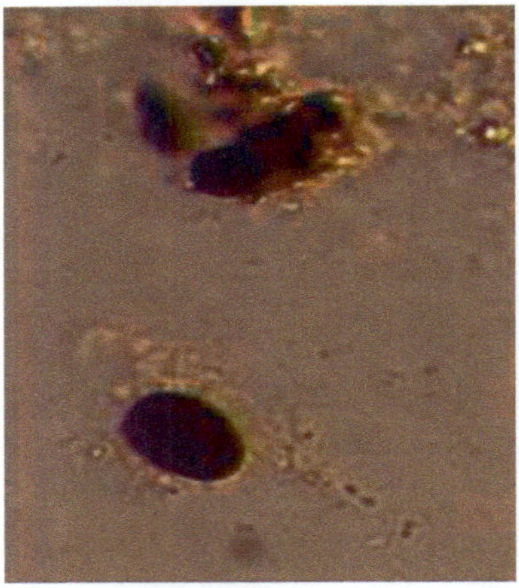

FIG. 5.19.6 Starch grains from *Syzygium cumini* bark powder.

References

[1] Government of India, Department of AYUSH, Ayurvedic Pharmacopoeia of India. Part I, vol. II, Ministry of Health and Welfare, Ghaziabad, 1999–2017, p. 60.

[2] M.G. Chauhan, A.P.G. Pillai, Microscopic Profile of Powdered Drugs Used in Indian System of Medicine, Bark Drugs, vol. 1, Institute of Ayurvedic Medicinal Plant Sciences, Jamnagar, Gujarat, India, 2005, p. 44.

Chapter 5.20

Zizyphus mauritiana

Zizyphus mauritiana Lam. Rhamnaceae

Common vernacular names: Assamese – Bagori, Bayur; Bengali – Kula; English – Jujube; Gujarati – Bor; Hindi – Desi Ber; Kannada – Boehannumara; Malayalam – Lanta; Marathi – Bor; Oriya – Borakali; Punjabi – Desi ber; Tamil – Ilandai; Telugu – Regi, Regu [1] (Fig. 5.20.1).

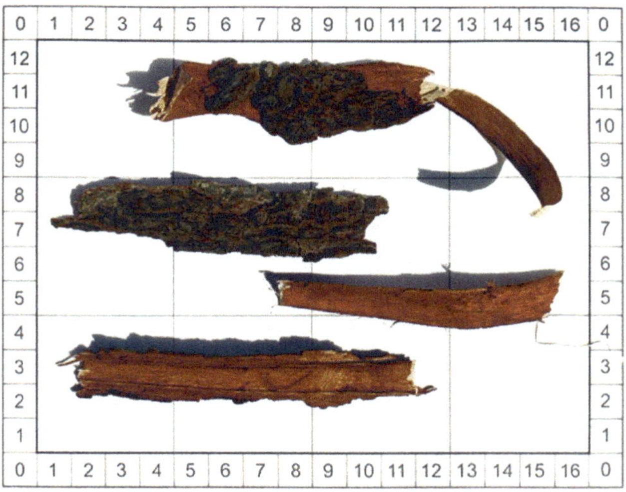

FIG. 5.20.1 *Zizyphus mauritiana* bark.

Macroscopic description: Shape – Flat, thick, pieces of variable length; **Outer bark** – Blackish grey, hard, rough due to deep furrows and fissures, exfoliating into irregular scales; **Inner bark** – Reddish brown fibrous zones; **Fracture** – Fibrous; **Odour** – Nil.

Major chemical constituents: Cyclopeptide alkaloids [1].

Traditional uses: Diarrhea, dysentery, Astringent, Rheumatism [1].

Powder microscopic description: Reddish brown powder showing abundant prismatic crystals; Oval to rectangular stone cells; Crystal fibre and absence of sclereids [2]. Microscopical features are not properly visible even after potassium chlorate treatment. The general as well as specific features of *Z*

Powdered Crude Drug Microscopy of Leaves and Barks. https://doi.org/10.1016/B978-0-12-818092-1.00024-5

.mauritiana bark powder are illustrated in Table 5.20.1 and Figs. 5.20.2A–B and 5.20.3 respectively.

TABLE 5.20.1 Specific as well as general characters of bark powder of *Z. mauritiana*.

	Specific features	General features
	Stone cells	**Calcium oxalate crystals**
Occurrence	Few	Prismatic
	Isolated	
Walls	Thin walled	
	Radiating pits	
Shape	Oval	
	Rectangular	
Lumen	Wide	

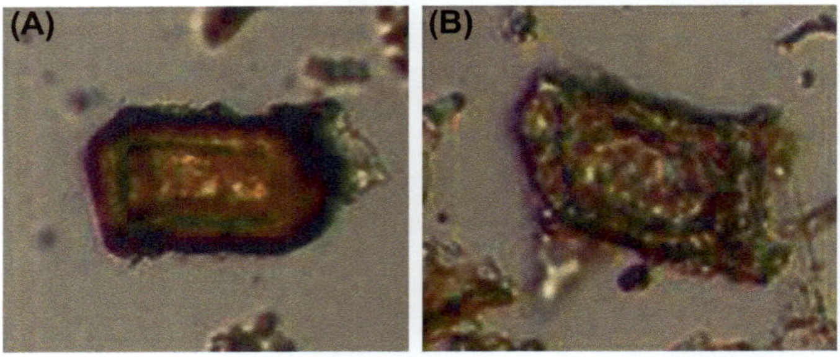

FIG. 5.20.2 (A–B) – Stone cells from *Zizyphus mauritiana* bark powder.

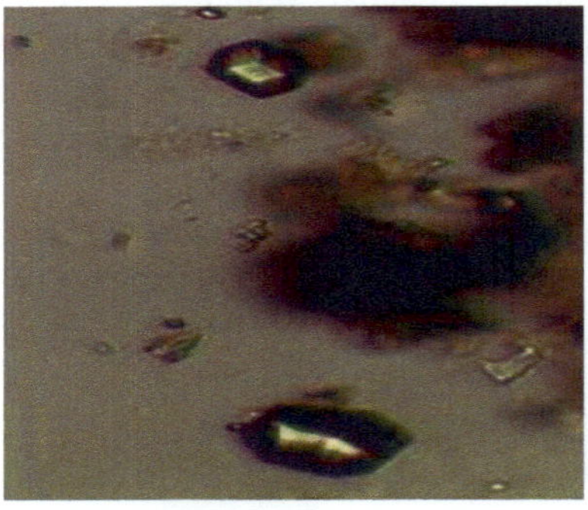

FIG. 5.20.3 Prismatic crystals from *Zizyphus mauritiana* bark powder.

References

[1] Government of India, Department of AYUSH, Ayurvedic Pharmacopoeia of India. Part I, vol. III, Ministry of Health and Welfare, Ghaziabad, 1999–2017, p. 97.

[2] M.G. Chauhan, A.P.G. Pillai, Microscopic Profile of Powdered Drugs Used in Indian System of Medicine, Bark Drugs, vol. 1, Institute of Ayurvedic Medicinal Plant Sciences, Jamnagar, Gujarat, India, 2005, p. 26.

Chapter 6

Microscopy/Monograph of powdered leaf drugs

The features selected for identification of powdered leaf samples includes – epidermis, stomata, trichomes, calcium oxalate crystals, fibres, cell contents, cystoliths, lamina, starch grains, tracheids, lactiferous canals, and xylem vessels which are further differentiated into various sub features. The below mentioned leaves are subjected to detailed powder microscopic evaluation using Motic microscope moticam 3.0 MP, AE 2000. The parameters selected for the study of morphology of leaf sample are – Shape, apex, base, margin, venation, petiole, colour and odour. The photographs are taken in a grid using Cannon 500D camera with 18-megapixel capacity. The leaves selected for macroscopic and powder microscopic examination are listed below:

6.1 *Acacia pennata* L. Willd.
6.2 *Aloe barbadensis* Mill.
6.3 *Aristolochia bracteata* Retz.
6.4 *Cannabis sativa* L.
6.5 *Chrysanthemum indicum* L.
6.6 *Cinnamomum tamala* (Buch. -Ham.) T.Nees & Eberm.
6.7 *Citrullus colocynthis* (L.) Schrad.
6.8 *Coccinia grandis* (L.) Voigt
6.9 *Coleus amboinicus* Lour.
6.10 *Gymnema sylvestre* (Retz.) R.Br. ex Sm.
6.11 *Jasminum officinale* L.
6.12 *Murraya koenigii* (L.) Spreng.
6.13 *Piper betle* L.
6.14 *Plantago lanceolata* L.
6.15 *Pongamia pinnata* (L.) Pierre
6.16 *Prosopis cineraria* (L.) Druce
6.17 *Punica granatum* L.
6.18 *Ricinus communis* L.
6.19 *Sesbania sesban* (L.) Merr.
6.20 *Sphaeranthus indicus* L.

Chapter 6.1

Acacia pennata

Acacia pennata L. Willd. Mimosaceae

Common vernacular names: Bengali – Kuchai; Gujarati – Khervelya; Hindi – Biswal, Latakhadira, Aazi Khair; Kannada – Siguri; Marathi – Aarai velyakhera; Oriya – Potadontari; Tamil – Iya kozhundu; Telugu – Karusakaya [1] (Fig. 6.1.1).

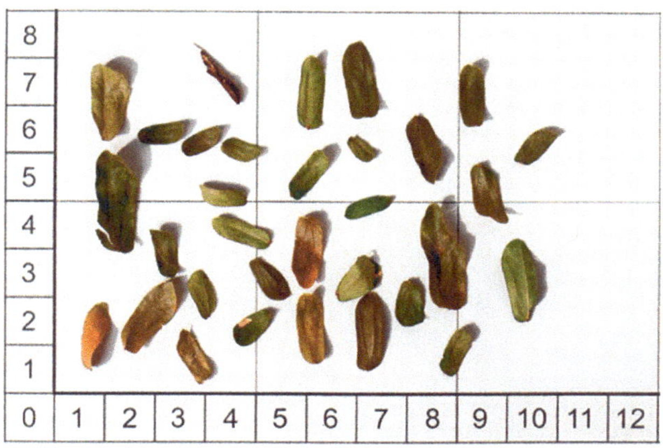

FIG. 6.1.1 *Acacia pennata* leaf.

Macroscopic description: Shape – Oblong, **Apex** – Obtuse to shortly acuminate, **Base** – Rounded, **Margin** – Entire, **Venation** – Lateral veins difficult to distinguish on fresh material, **Petiole** – Very short, **Colour** – Light green to yellowish green, **Odour** – Characteristic.

Major chemical constituents: Taepeenin D, Octadecadienoic, octadecanoic, palmitic and pentadecanoic acids, lupeol, a-spinasterol, sitosterol and tannins [2].

Traditional uses: Fever, Disorder of blood, Digestive impairment [1].

Powder microscopic description: Pale greenish yellow powder showing fragments of upper and lower epidermis, wavy walled embedded with paracytic stomata, prismatic crystals of calcium oxalate and starch grains scattered

Powdered Crude Drug Microscopy of Leaves and Barks. https://doi.org/10.1016/B978-0-12-818092-1.00025-7

as such. Simple, unicellular covering trichome along with stellate trichomes are also present. The general as well as specific features of *A. pennata* leaf powder are illustrated in Table 6.1.1 and Figs. 6.1.2–6.1.4 respectively.

TABLE 6.1.1 Specific as well as general characters of leaf powder of *Acacia pennata*.

Features	
Epidermis	**Trichomes**
Wavy	Simple
Anticlinical	Unicellular
	Covering
Stomata	Stellate
Paracytic	**Calcium oxalate crystals**
	Prismatic

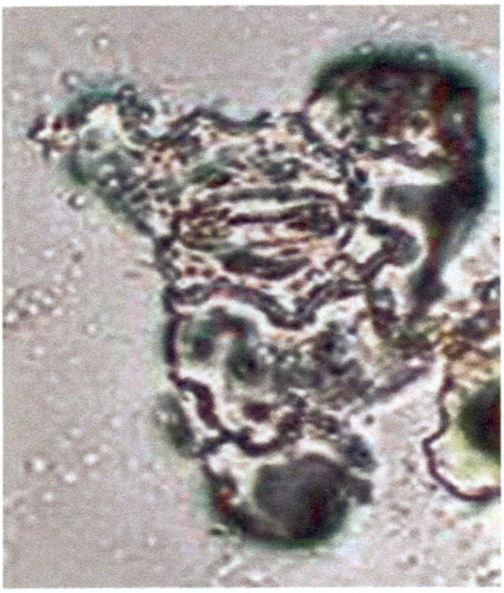

FIG. 6.1.2 Stomata from *Acacia pennata* leaf powder.

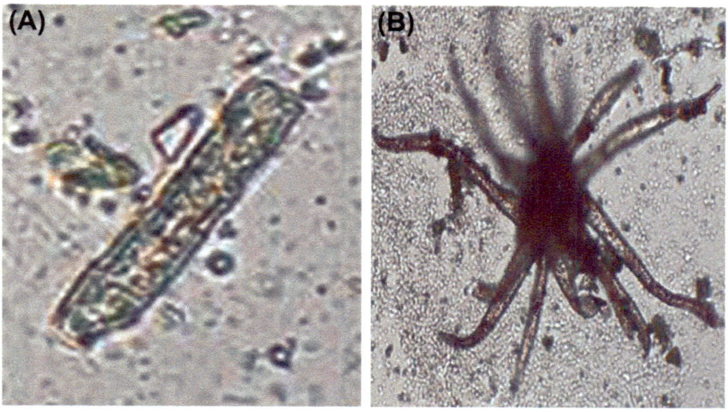

FIG. 6.1.3 (A–B) – Covering and stellate trichomes from *Acacia pennata* leaf powder.

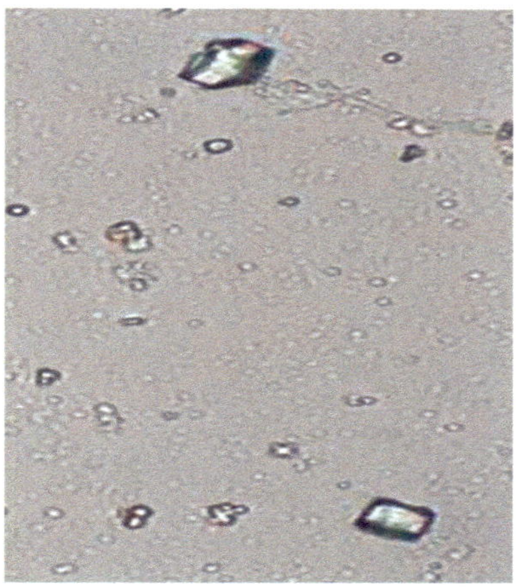

FIG. 6.1.4 Prismatic crystals from *Acacia pennata* leaf powder.

References

[1] Government of India, Department of AYUSH, Ayurvedic Pharmacopoeia of India. Part I, vol. VI, Ministry of Health and Welfare, Ghaziabad, 1999–2017, p. XL.

[2] A.K. Gupta, N. Tandon, M. Sharma, Quality Standards of Indian Medicinal Plants, vol. 11, Medicinal Plant Unit, ICMR, New Delhi, India, 2008, p. 03.

Chapter 6.2

Aloe barbadensis

Aloe barbadensis Mill Liliaceae

Common vernacular names: Assamese – Musabhar, Machambar; Bengali – Ghritakalmi; English – Indian Aloe; Gujarati – Eliyo, Eariyo; Hindi – Musabhar, Elva; Kannada – Karibola, Lolesara satva, Lovalsara, Lolesara; Kashmiri – Musabbar, Siber; Malayalam – Chenninayakam; Marathi – Korphad; Oriya – Musabara; Punjabi – Kalasohaga, Mussabar, Alua; Tamil – Kattazhi, Satthukkathazhai; Telugu – Musambaram [1] (Fig. 6.2.1).

FIG. 6.2.1 *Aloe barbadensis* leaf.

Macroscopic description: Shape – Flat or slightly concave on the upper surface and strongly rounded on the lower; **Apex** – Tapering with a strong spine at the apex; **Base** – Thick, colourless, mucilaginous mass at the base; **Margin** – Smaller spines at the margin; **Colour** – Dark green with white blotches at places; **Odour** – Unpleasant.

Powdered Crude Drug Microscopy of Leaves and Barks. https://doi.org/10.1016/B978-0-12-818092-1.00026-9

Major chemical constituents: Anthraquinone glycosides [1].

Traditional uses: Skin diseases and purgative [1].

Powder microscopic description: Powder when mounted in water and observed under microscope shows very large number of crystalline needles, yellowish brown to chocolate coloured particles and straight epidermal cells embedded with minute stomata which was not very clearly visible. The general as well as specific features of *A. barbadensis* leaf powder are illustrated in Table 6.2.1 and Figs. 6.2.2–6.2.6 respectively.

TABLE 6.2.1 Specific as well as general characters of leaf powder of *Aloe barbadensis*.

Features	
Epidermis	**Stomata**
Straight	Sunken
Anticlinical	
	Cell contents
Calcium oxalate crystals	Brownish matter
Idioblast	
Needles	

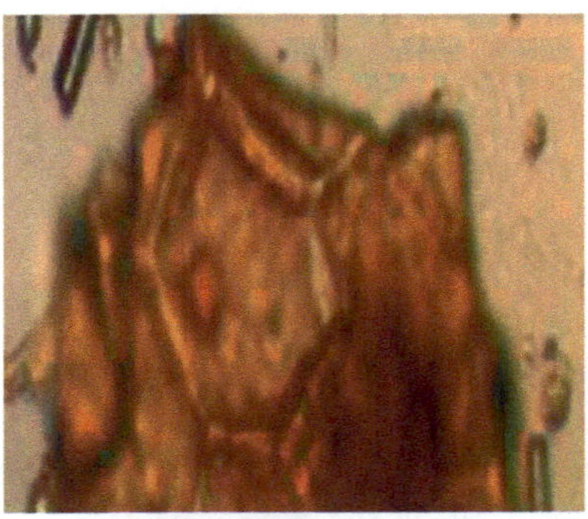

FIG. 6.2.2 Epidermal cells present in *Aloe barbadensis* leaf powder.

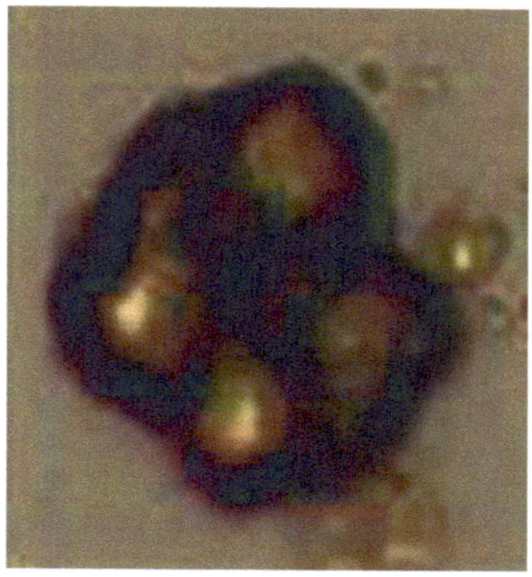

FIG. 6.2.3 Idioblast present in *Aloe barbadensis* leaf powder.

FIG. 6.2.4 Crystalline needles present in *Aloe barbadensis* leaf powder.

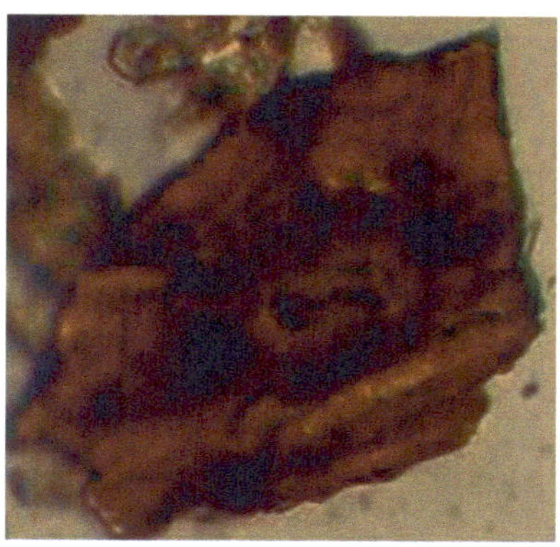

FIG. 6.2.5 Sunken stomata present in *Aloe barbadensis* leaf powder.

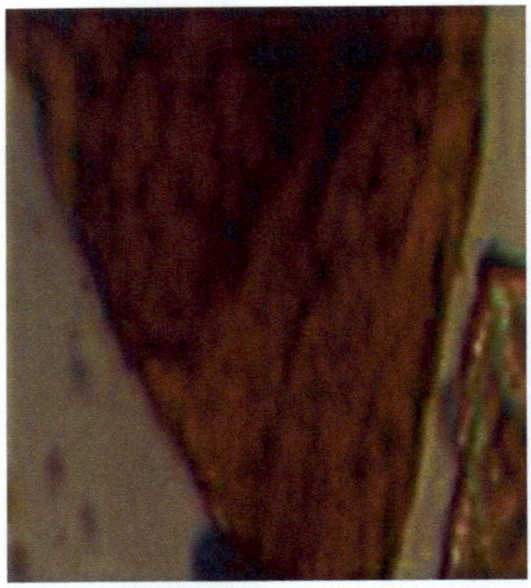

FIG. 6.2.6 Brownish matter present in *Aloe barbadensis* leaf powder.

Reference

[1] Government of India, Department of AYUSH, Ayurvedic Pharmacopoeia of India. Part I, vol. I, Ministry of Health and Welfare, Ghaziabad, 1999–2017, p. 81.

Chapter 6.3

Aristolochia bracteata

Aristolochia bracteata Retz. Aristolochiaceae

Common vernacular names: Bengali – Kiramar; English – Bracteated birthwort; Gujarati – Kidaamaari; Hindi – Kitmaar, Kiramar, Kitmaari, Kidaamaari; Kannada – Kathhekirubanagida; Malayalam – Aduthinapalai, Atu-tinlap; Marathi – Kidaamaari, Kidemaar; Oriya – Paaniri; Punjabi – Kitamar; Tamil – Aadu-tinna-paalai; Telugu – Gadida gadapa, Tella iswari [1] (Fig. 6.3.1).

FIG. 6.3.1 *Aristolochia bracteata* leaf.

Macroscopic description: Shape – Ovate to cordate; **Apex** – Acuminate; **Base** – Cordate; **Margin** – Entire; **Venation** – Finely reticulately veined; **Petiole** – Cylindrical, slightly grooved on the upper surface; **Colour** – Dark green when fresh glabrous above and glaucous beneath; **Odour** – Feebly aromatic when crushed but not characteristic.

Major chemical constituents: Aristolochic acid, ceryl alcohol, β-sitosterol [2].

Powdered Crude Drug Microscopy of Leaves and Barks. https://doi.org/10.1016/B978-0-12-818092-1.00027-0

Traditional uses: Stomachic, uterine tonic, anthelminthic, febrifuge [1].

Powder microscopic description: Greyish green powder showing fragments of upper and lower epidermis with straight or slightly sinous walls embedded with anisocytic or anomocytic stomata; Longitudinally cut fragments of spiral xylem vessels and few, simple, unicellular, uniseriate, covering trichomes. The general as well as specific features of *A. bracteata* leaf powder are illustrated in Table 6.3.1 and Fig. 6.3.2–6.3.4 respectively.

TABLE 6.3.1 Specific as well as general characters of leaf powder of *Aristolochia bracteata*.

Features	
Epidermis	**Trichomes**
Straight	Few
Slightly sinous	Simple
Anticlinical	Unicellular
Stomata	Uniseriate
Anisocytic	Covering
Anomocytic	Pointed apex
Xylem vessels	Curved
Spiral	

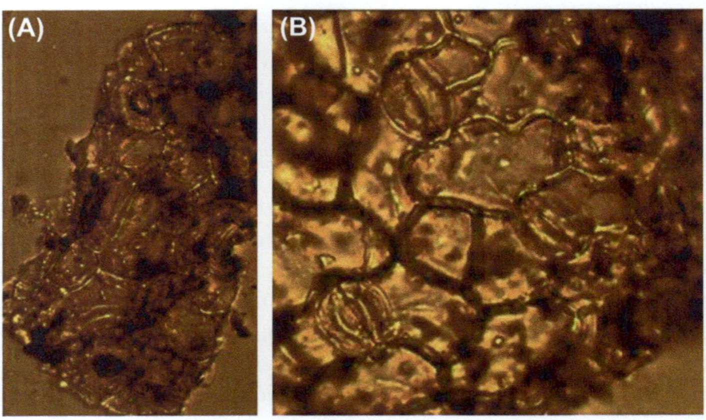

FIG. 6.3.2 (A–B) – Different types of stomata present in *Aristolochia bracteata* leaf powder.

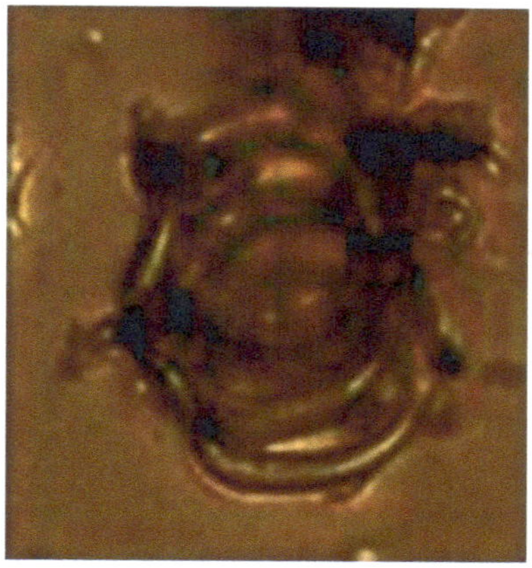

FIG. 6.3.3 Spiral xylem vessel present in *Aristolochia bracteata* leaf powder.

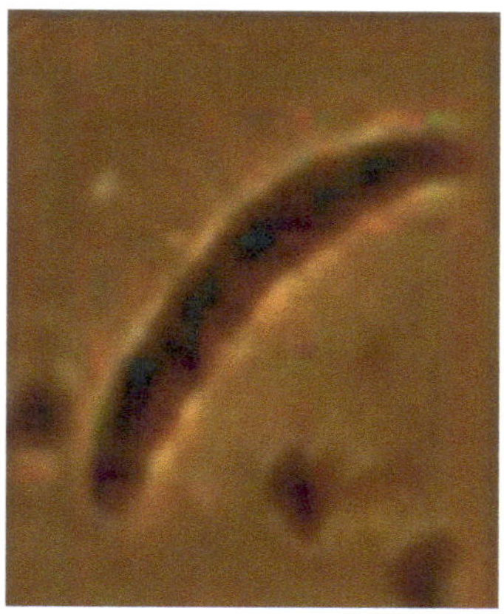

FIG. 6.3.4 Covering trichome present in *Aristolochia bracteata* leaf powder.

References

[1] Government of India, Department of AYUSH, Ayurvedic Pharmacopoeia of India. Part I, vol. VI, Ministry of Health and Welfare, Ghaziabad, 1999–2017, p. CXXVI.

[2] A.K. Gupta, N. Tandon, M. Sharma, Quality Standards of Indian Medicinal Plants, vol. 5, Medicinal Plant Unit, ICMR, New Delhi, India, 2008, p. 85.

Chapter 6.4

Cannabis sativa

Cannabis sativa L. Cannabinaceae

Common vernacular names: Assamese – Bhan, Bhang; Bengali – Bhang, Sidhi; English – Indian Hemp; Gujarati – Bhang; Hindi – Bhaang, Bhanga; Kannada – Bhangigida, Ganjagida; Kashmiri – Pang, Bangi; Malayalam – Kanchavu; Marathi – Bhang, Ganja; Oriya – Bhanga, Ganjei; Punjabi – Bhang; Tamil – Ganja; Telugu – Ganjayi [1] (Fig. 6.4.1).

FIG. 6.4.1 *Cannabis sativa* leaf.

Macroscopic description: Shape – Palmately compound leaflets, lanceolate; **Apex** – Acuminate; **Base** – Narrow; **Margin** – Serrate; **Venation** – Pinnate; **Petiole** – Little stalk that holds the main blade of the leaf; **Colour** – Dark green and rough, lower surface pale; **Odour** – Strong and characteristic.

Major chemical constituents: Resin (Cannabinols, particularly tetrahydrocannabinol) [1].

Traditional uses: Hypnotic and sedative [1].

Powdered Crude Drug Microscopy of Leaves and Barks. https://doi.org/10.1016/B978-0-12-818092-1.00028-2

Powder microscopic description: Dark green colour powder showing plenty of simple, covering, unicellular and conical trichomes along with few glandular trichomes with broken head and detached from stalk; Prismatic and rosette crystals of calcium oxalate; Longitudinally cut fragments of midrib embedded with resin ducts filled with brownish contents and fragments of spiral xylem vessels [2]. The general as well as specific features of *C. sativa* leaf powder are illustrated in Table 6.4.1 and Figs. 6.4.2–6.4.6A–D respectively.

TABLE 6.4.1 Specific as well as general characters of leaf powder of *Cannabis sativa.*

Features	
Calcium oxalate crystals	**Trichomes**
Prismatic	Abundant
Rosette	Simple
	Covering
Cell contents	Glandular
Resin ducts	Unicellular
Brownish matter	Uniseriate
	Short stalk unicellular
Xylem vessels	Conical
Spiral	Cystolithic
	Pointed apex
	Curved

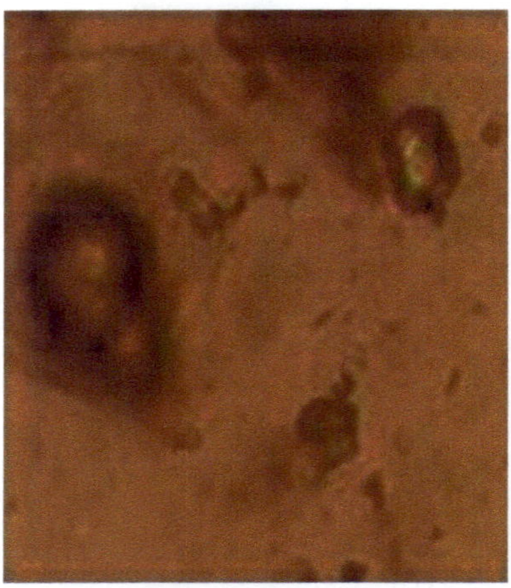

FIG. 6.4.2 Prismatic crystals present in *Cannabis sativa* leaf powder.

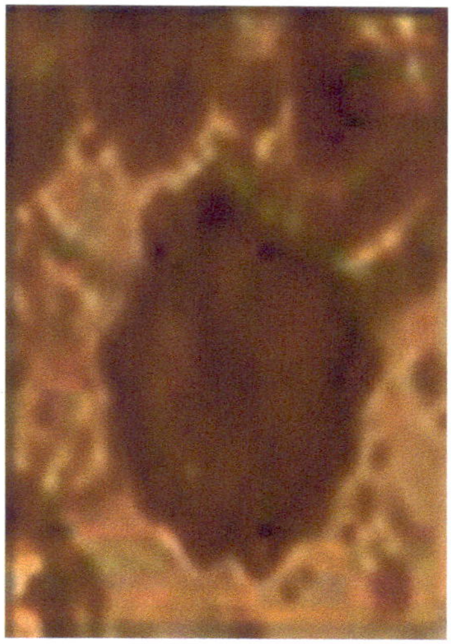

FIG. 6.4.3 Rosette crystals present in *Cannabis sativa* leaf powder.

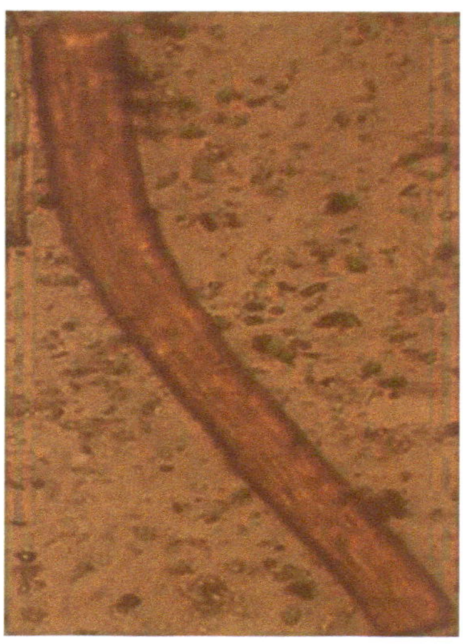

FIG. 6.4.4 Resin duct present in *Cannabis sativa* leaf powder.

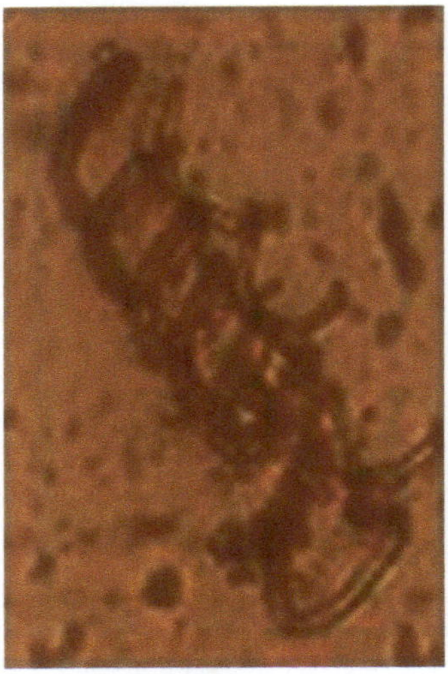

FIG. 6.4.5 Spiral xylem vessel present in *Cannabis sativa* leaf powder.

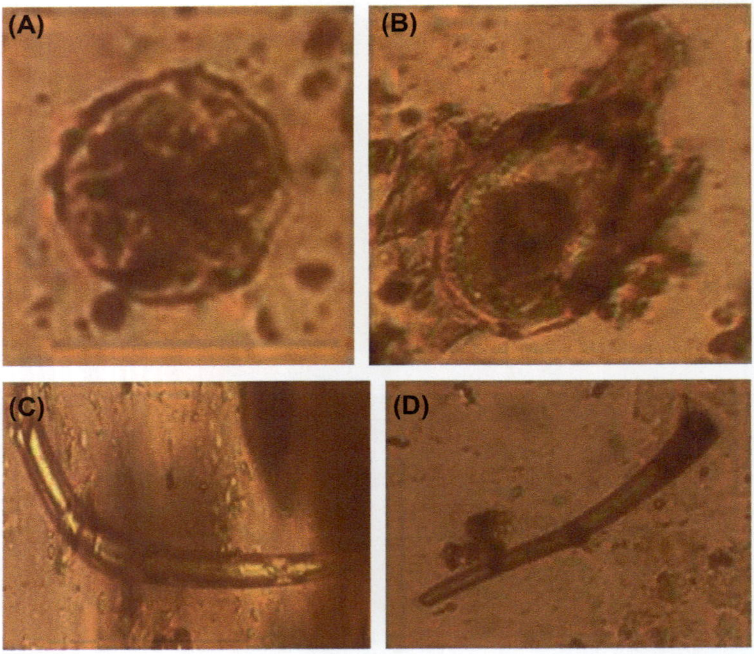

FIG. 6.4.6 (A–D) – Trichomes of various types present in *Cannabis sativa* leaf powder.

References

[1] A.K. Gupta, N. Tandon, M. Sharma, Quality Standards of Indian Medicinal Plants, vol. 4, Medicinal Plant Unit, ICMR, New Delhi, India, 2008, p. 46.
[2] M.G. Chauhan, A.P.G. Pillai, Microscopic Profile of Powdered Drugs Used in Indian System of Medicine, Leaf Drugs, vol. 2, Institute of Ayurvedic Medicinal Plant Sciences, Jamnagar, Gujarat, India, 2007, p. 22.

Chapter 6.5

Chrysanthemum indicum

Chrysanthemum indicum L. Asteraceae

Common vernacular names: Bengali – Chandramukhi; English – Chrysanthemum; Gujarati – Guldaaudi; Hindi – Guldaaudi; Kannada – Chamanti; Marathi – Shevanti; Punjabi – Bagaura; Tamil – Chamanti; Telugu – Chamanti [1] (Fig. 6.5.1).

FIG. 6.5.1 *Chrysanthemum indicum* leaf.

Macroscopic description: Shape – Pinnatipartite; **Apex** – Acute; **Base** – Truncate; **Margin** – Lacerate; **Venation** – Palmately reticulated veins; **Petiole** – Short, slightly winged; **Colour** – Dark green on upper surface and dull green on lower surface; **Odour** – Aromatic.

Major chemical constituents: Sesquiterpene lactones – angeloylcumambrin B, arteglasin A and angleloylajadin, Chrysanthenone and chrysanthenine glucoside. Aerial parts also contain lignans sesamin and fargesin and flavonoid penduletin [1].

Traditional uses: Migraine, Ulcer in mouth and skin diseases [1].

Powder microscopic description: Greenish yellow powder showing fragments of upper and lower epidermis, slightly sinous thick walled

Powdered Crude Drug Microscopy of Leaves and Barks. https://doi.org/10.1016/B978-0-12-818092-1.00029-4

embedded with diacytic stomata; Plenty of trichomes of various shapes and sizes – simple, covering trichomes with oval to globular head or pointed apex, straight or slightly bent, unicellular to multicellular, abundant shaggy wooly trichomes and small prismatic crystals of calcium oxalate. The general as well as specific features of *C. indicum* leaf powder are illustrated in Table 6.5.1 and Figs. 6.5.2–6.5.4A–E respectively.

TABLE 6.5.1 Specific as well as general characters of leaf powder of *Chrysanthemum indicum*.

Features	
Epidermis	**Trichomes**
Thick	Abundant
Slightly sinous walled	Simple
	Covering
Stomata	Unicellular
Diacytic	Multicellular
	Uniseriate
	Shaggy wooly trichomes
Calcium oxalate crystals	Straight
Prismatic	Globular head
	Pointed apex
	Curved
	Bent
	Long

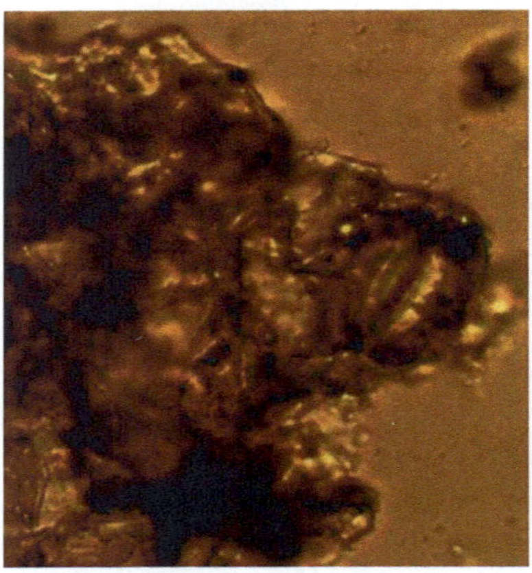

FIG. 6.5.2 Diacytic stomata present in *Chrysanthemum indicum* leaf powder.

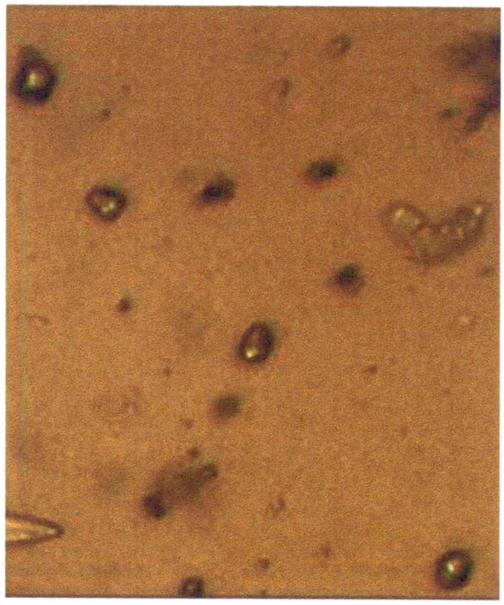

FIG. 6.5.3 Prismatic crystals present in *Chrysanthemum indicum* leaf powder.

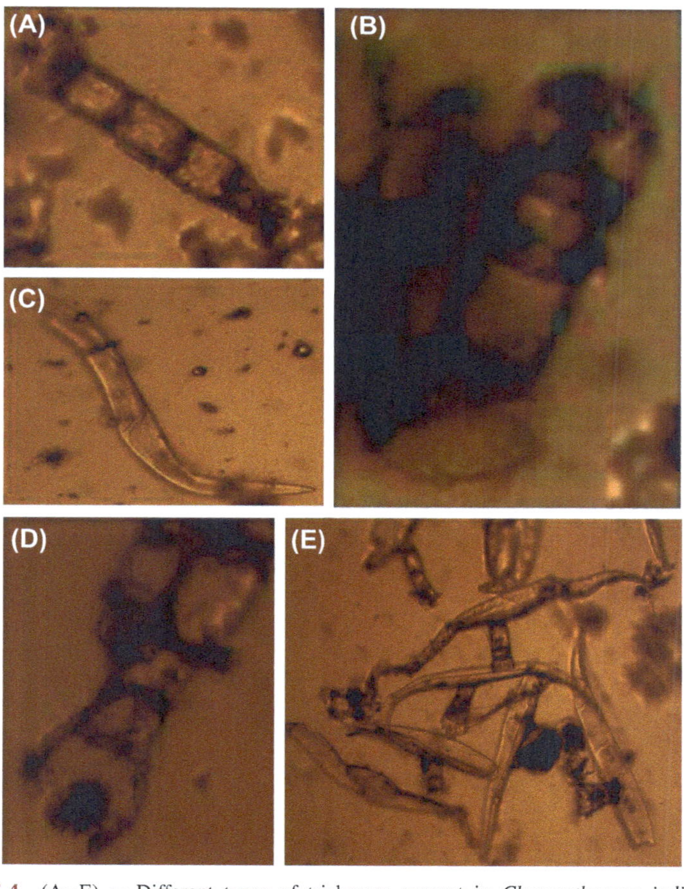

FIG. 6.5.4 (A–E) – Different types of trichomes present in *Chrysanthemum indicum* leaf powder.

Reference

[1] Government of India, Department of AYUSH, Ayurvedic Pharmacopoeia of India. Part I, vol. VI, Ministry of Health and Welfare, Ghaziabad, 1999–2017, p. CI.

Chapter 6.6

Cinnamomum tamala

Cinnamomum tamala (Buch.-Ham.) T.Nees & Eberm. Lauraceae

Common vernacular names: Assamese – Tejpat, Mahpat; Bengali – Tejpatra, Tejpata; English – Indian Cinnamon; Gujarati – Tamala patra, Develee; Hindi – Tejpatra; Kannada – Tamalapatra, Dalchini Ele; Kashmiri – Dalchini pan, Tajpatra; Malayalam – Karuvapatta patram; Marathi – Tamalpatra; Oriya – Tejapatra; Punjabi – Tajpater; Sanskrit – Patra, Variga, Coca; Tamil – Lavangapatri; Telugu – Akupatri [1] (Fig. 6.6.1).

FIG. 6.6.1 *Cinnamomum tamala* leaf.

Macroscopic description: Shape – Lanceolate; **Apex** – Acute to acuminate; **Base** – Acute; **Margin** – Entire; **Venation** – Palmately parallel, converging, 3 converging nerves from base to apex; **Petiole** – Short; **Colour** – Olive green in colour; **Odour** – Aromatic.

Powdered Crude Drug Microscopy of Leaves and Barks. https://doi.org/10.1016/B978-0-12-818092-1.00030-0

Major chemical constituents: Essential oils (d-α phellandrene and eugenol) [1].

Traditional uses: Piles, Anoerexia [1].

Powder microscopic description: Olive green colour powder showing epidermis with sinous anticlinical walls, embedded with paracytic stomata; Simple, unicellular thick walled trichomes some being very short, while other moderately longer in height; Plenty of stone cells and sclereids of various shape and sizes, wide lumened and plenty of prismatic and rod shaped crystals of calcium oxalate [2,3]. The general as well as specific features of *C. tamala* leaf powder are illustrated in Table 6.6.1 and Figs. 6.6.2–6.6.5A–B respectively.

TABLE 6.6.1 Specific as well as general characters of leaf powder of *Cinnamomum tamala.*

Features		
Epidermis	**Trichomes**	**Calcium oxalate crystals**
Slightly sinous	Abundant	Prismatic
Anticlinical	Simple	
Thick walled	Covering	
	Unicellular	**Stone cells and sclereids**
Stomata	Thick walled	Lignified
Paracytic	Short	Wide lumen
	Long	Oval
	Curved	Triangular
	Pointed apex	

FIG. 6.6.2 Paracytic stomata present in *Cinnamomum tamala* leaf powder.

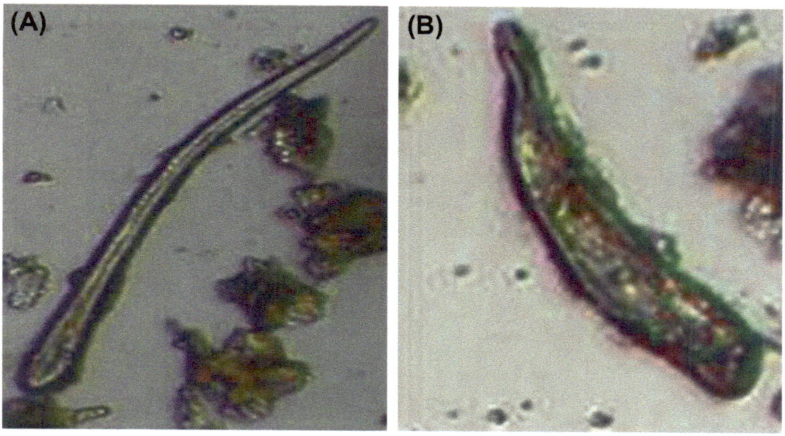

FIG. 6.6.3 (A–B) – Different types of trichomes present in *Cinnamomum tamala* leaf powder.

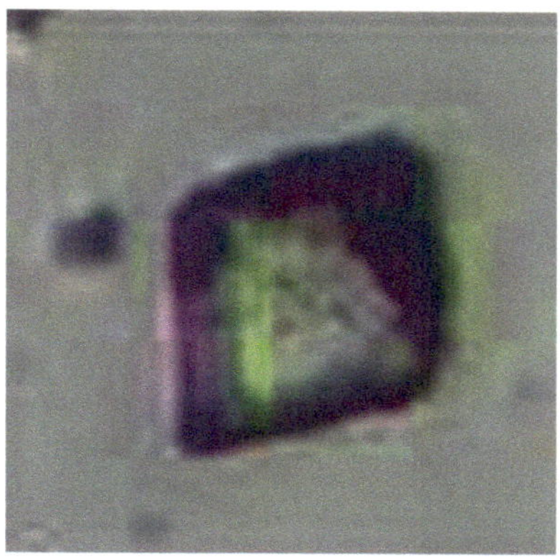

FIG. 6.6.4 Prismatic crystals present in *Cinnamomum tamala* leaf powder.

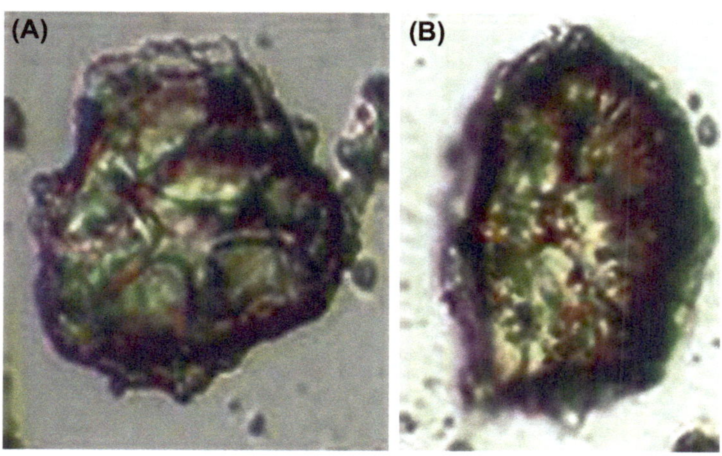

FIG. 6.6.5 (A–B) – Stone cells and sclereids present in *Cinnamomum tamala* leaf powder.

References

[1] Government of India, Department of AYUSH, Ayurvedic Pharmacopoeia of India. Part I, vol. I, Ministry of Health and Welfare, Ghaziabad, 1999–2017, p. 154.

[2] A.K. Gupta, N. Tandon, M. Sharma, Quality Standards of Indian Medicinal Plants, vol. 3, Medicinal Plant Unit, ICMR, New Delhi, India, 2008, p. 149.

[3] M.G. Chauhan, A.P.G. Pillai, Microscopic Profile of Powdered Drugs Used in Indian System of Medicine, Leaf Drugs, vol. 2, Institute of Ayurvedic Medicinal Plant Sciences, Jamnagar, Gujarat, India, 2007, p. 154.

Chapter 6.7

Citrullus colocynthis

Citrullus colocynthis (L.) Schrad. Cucurbitaceae

Common vernacular names: Assamese – Nantiyah; Bengali – Rahhalasa, Makhal; English – Colocynth, Bitter Apple; Gujarati – Indrayana, Indrayanoa, Insbak; Hindi – Indrayana; Kannada – Havumekke Kayi, Havamikke; Malayalam – Kattu vellari, Kadu Indrayan, Peykommuti; Marathi – Indrayana, Kodu indrayan; Oriya – Gothkakudi, Mahakal; Punjabi – Tumma, Jamtumma; Tamil – Peyakkumutti, Peytumatti, Peyththumatti, Peykhumutti, Verittumatti; Telugu – Chedupuchcha [1] (Fig. 6.7.1).

FIG. 6.7.1 *Citrullus colocynthis* leaf.

Macroscopic description: Shape – Simple, pinnatifid in outline; **Apex** – Subacute to obtuse; **Base** – Sagittate; **Margin** – Minutely toothed; **Venation** – Pinnatifid; **Petiole** – Short, cylindrical; **Colour** – Upper dark and lower pale green; **Odour** – Characteristic.

Major chemical constituents: Quercetin, Kaempferol, Cucurbitacins B, P, L and I, glucosides of cucurbitacin Band E, a-elaterin 2-Dglucopyranoside [2].

Powdered Crude Drug Microscopy of Leaves and Barks. https://doi.org/10.1016/B978-0-12-818092-1.00031-2

Traditional uses: Dermatopathies [2].

Powder microscopic description: Dark green colour powder showing fragments of upper and lower epidermis, slightly sinous walled embedded with anomocytic stomata, oil glands and numerous oil globules; Longitudinally cut fragments of spiral helical vessels and devoid of trichomes and calcium oxalate crystals. The general as well as specific features of *C. colocynthis* leaf powder are illustrated in Table 6.7.1 and Figs. 6.7.2–6.7.4 respectively.

TABLE 6.7.1 Specific as well as general characters of leaf powder of *Citrullus colocynthis.*

Features	
Epidermis	**Stomata**
Slightly sinous	Anomocytic
Anticlinical	
Oil glands	**Cell contents**
Xylem Vessels	Volatile oil
Spiral	

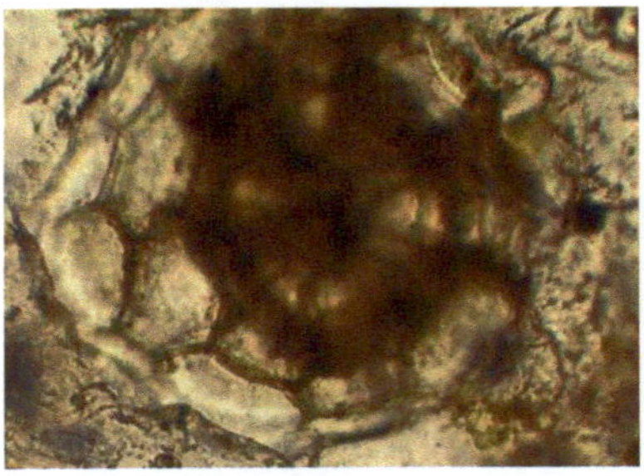

FIG. 6.7.2 Oil gland present in *Citrullus colocynthis* leaf powder.

FIG. 6.7.3 Spiral xylem vessel present in *Citrullus colocynthis* leaf powder.

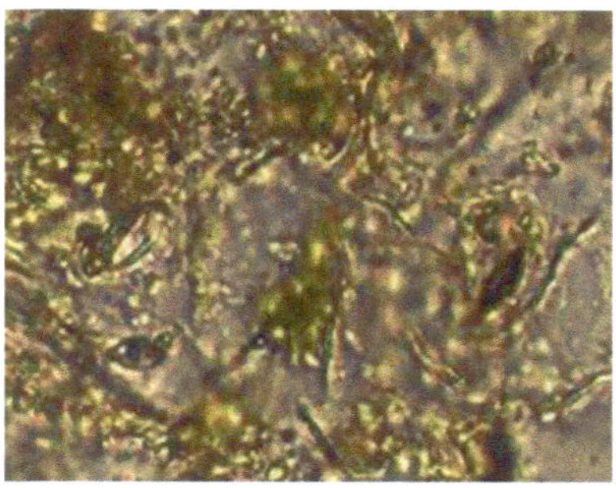

FIG. 6.7.4 Anomocytic stomata present in *Citrullus colocynthis* leaf powder.

References

[1] Government of India, Department of AYUSH, Ayurvedic Pharmacopoeia of India. Part I, vol. II, Ministry of Health and Welfare, Ghaziabad, 1999–2017, p. 55.

[2] A.K. Gupta, N. Tandon, M. Sharma, Quality Standards of Indian Medicinal Plants, vol. 11, Medicinal Plant Unit, ICMR, New Delhi, India, 2008, p. 143.

Chapter 6.8

Coccinia grandis

Coccinia grandis (L.) Voigt *Cucurbitaceae*

Common vernacular names: Assamese – Kanabhaturi; Bengali – Tela Kuccha, Bimbu; English – Ivy gourd; Gujarati – Gholam, Ghilodi, Tindoran, Kadavi Ghilodi; Hindi – Kunduru, Kunru; Kannada – Tonde balli; Malayalam – Koval, Kova, Nallakova; Marathi – Tondlee; Oriya – Kainchi kakudi, Bano Kundri; Punjabi – Kunduru, Kunduri; Tamil – Kovai; Telugu – Donda tige [1] (Fig. 6.8.1).

FIG. 6.8.1 *Coccinia grandis* leaf.

Macroscopic description: Shape – Palmatifid; **Apex** – Acute; **Base** – Cordate; **Margin** – Sinuate and toothed; **Venation** – Palmately reticulate with five main veins; **Petiole** – Long and cylindrical; **Colour** – Bright green above and paler beneath; **Odour** – Characteristic.

Powdered Crude Drug Microscopy of Leaves and Barks. https://doi.org/10.1016/B978-0-12-818092-1.00032-4

Major chemical constituents: Cephalandrine A and B, b-sitosterol and triacontane [1].

Traditional uses: Jaundice, Diabetes mellitus, Dysuria [1].

Powder microscopic description: Bright green colour powder showing fragments of upper and lower epidermis embedded with anomocytic stomata; few prismatic crystals of calcium oxalate; longitudinally cut fibre fragments and plenty of covering trichomes, uniseriate to biseriate, unicellular to multicellular, with rounded or pointed apex. The general as well as specific features of *C. grandis* leaf powder are illustrated in Table 6.8.1 and Figs. 6.8.2–6.8.5A–E respectively.

TABLE 6.8.1 Specific as well as general characters of leaf powder of *Coccinia grandis*.

Features	
Stomata	**Trichomes**
Anomocytic	Abundant
	Covering
Calcium oxalate crystal	Unicellular
Prismatic	Thick walled
	Multicellular
Fibres	Uniseriate
Isolated	Biseriate
Thick walled	Straight
Longitudinally cut	Pedestial base
	Pointed apex
	Rounded head

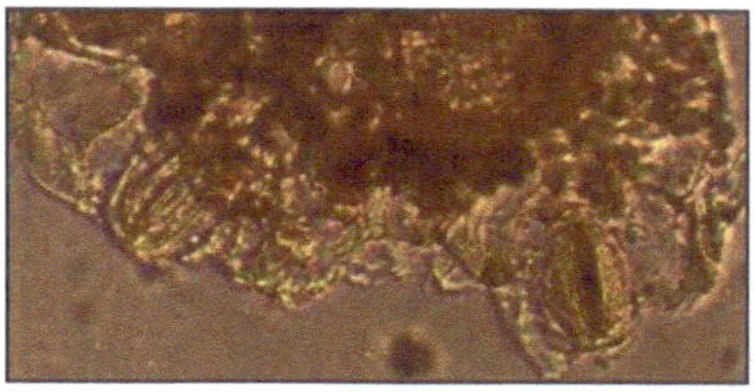

FIG. 6.8.2 Anomocytic stomata present in *Coccinia grandis* leaf powder.

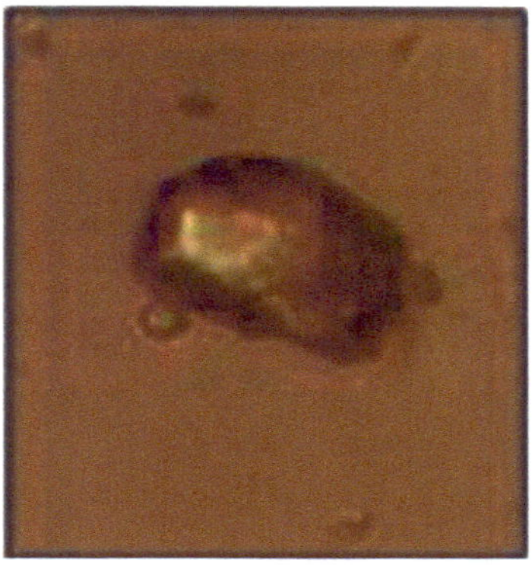

FIG. 6.8.3 Prismatic crystals present in *Coccinia grandis* leaf powder.

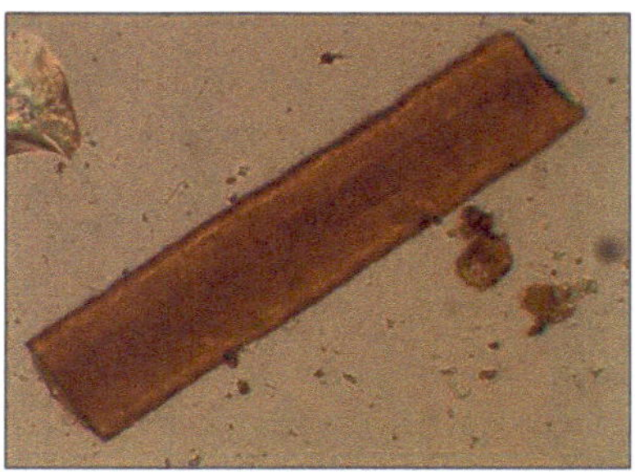

FIG. 6.8.4 Fibre fragment present in *Coccinia grandis* leaf powder.

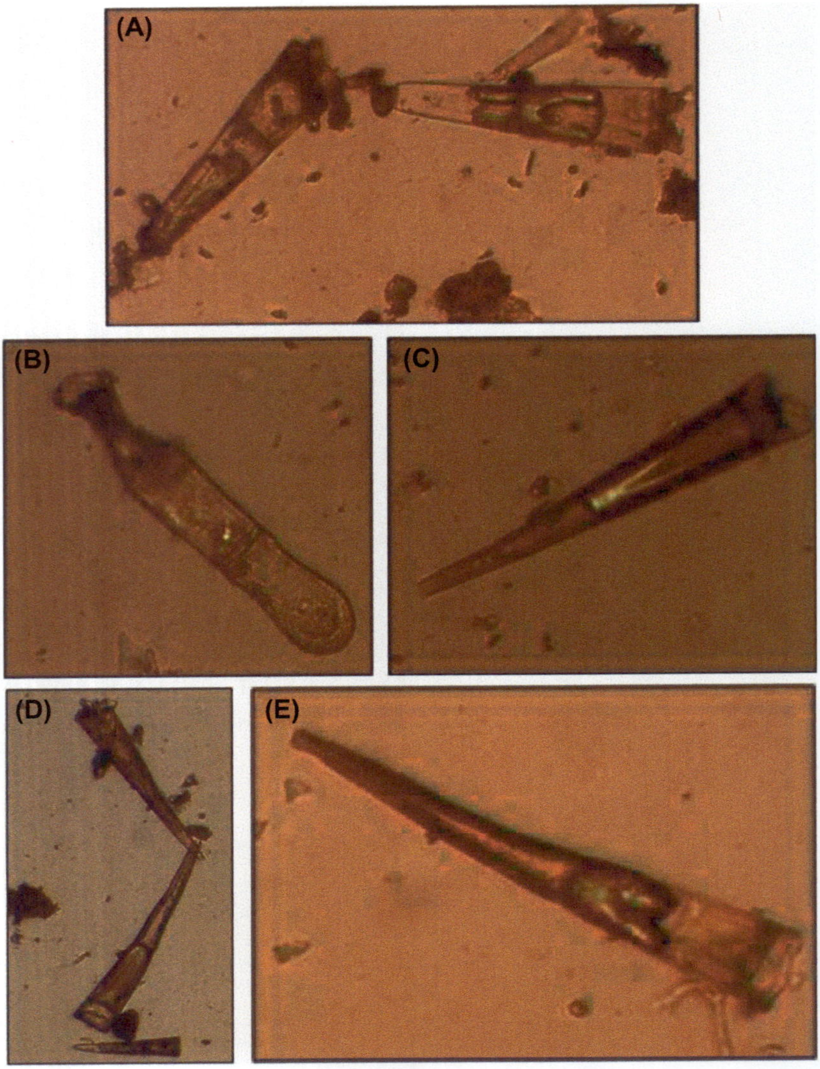

FIG. 6.8.5 (A – E) Different types of trichomes present in *Coccinia grandis* leaf powder.

Reference

[1] Government of India, Department of AYUSH, Ayurvedic Pharmacopoeia of India. Part I, vol. VI, Ministry of Health and Welfare, Ghaziabad, 1999–2017, p. LIX.

Chapter 6.9

Nerium indicum

Nerium indicum Mill. Apocynaceae

Common vernacular names: Assamese – Karbira, Karavi, Karvir; Bengali – Karavi, Kalkephul; English – Indian Oleander; Gujarati – Kanera, Karena, Karen; Hindi – Kaner; Kannada – Kanagalu, Kanagile; Kashmiri – Gandeela, Gandula; Malayalam – Kanave eram, Arali, Kattalari; Marathi – Kanher; Oriya – Kaniara, Kaniar; Punjabi – Kaner; Tamil – Arali, Alari, Aatrulari; Telugu – Ganneru [1] (Fig. 6.9.1).

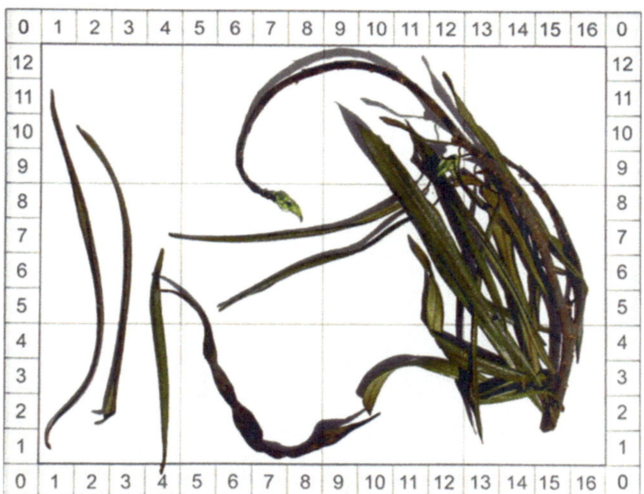

FIG. 6.9.1 *Nerium indicum* leaf.

Macroscopic description: Shape – Linear lanceolate; **Apex** – Acute to mucronate; **Base** – Cuneate; **Margin** – Entire; **Venation** – Unicostate reticulate; **Petiole** – Flattened ventrally; **Colour** – Upper surface dark green, lower paler in colour; **Odour** – Characteristic.

Major chemical constituents: Oleandrin and Ursolic acid [2].

Traditional uses: Heart diseases [1].

Powdered Crude Drug Microscopy of Leaves and Barks. https://doi.org/10.1016/B978-0-12-818092-1.00033-6

Powder microscopic description: Green colour powder showing fragments of upper and lower epidermis exhibited few minute sunken stomata; Simple, unicellular, straight or bent, covering trichomes and cluster crystals of calcium oxalate [3]. The general as well as specific features of *N. indicum* leaf powder are illustrated in Table 6.9.1 and Figs. 6.9.2 and 6.9.3 respectively.

TABLE 6.9.1 Specific as well as general characters of leaf powder of *Nerium indicum*.

Features	
Epidermis	**Trichomes**
Straight	Few
Anticlinical	Simple
Stomata	Unicellular
Sunken	Bent
Calcium oxalate crystals	Pointed apex
Cluster	Covering

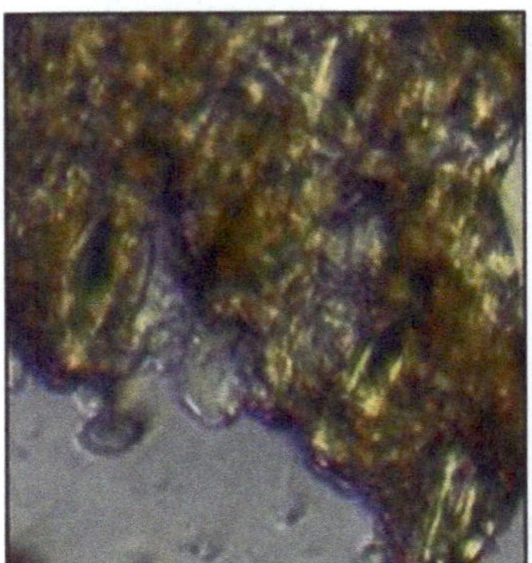

FIG. 6.9.2 Sunken stomata present in *Nerium indicum* leaf powder.

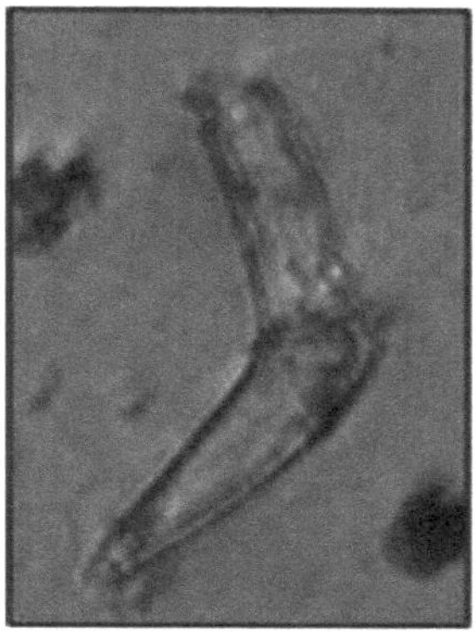

FIG. 6.9.3 Covering trichome present in *Nerium indicum* leaf powder.

References

[1] Government of India, Department of AYUSH, Ayurvedic Pharmacopoeia of India. Part I, vol. I, Ministry of Health and Welfare, Ghaziabad, 1999–2017, p. 85.

[2] A.K. Gupta, N. Tandon, M. Sharma, Quality Standards of Indian Medicinal Plants, vol. 11, Medicinal Plant Unit, ICMR, New Delhi, India, 2008, p. 271.

[3] M.G. Chauhan, A.P.G. Pillai, Microscopic Profile of Powdered Drugs Used in Indian System of Medicine, Leaf Drugs, vol. 2, Institute of Ayurvedic Medicinal Plant Sciences, Jamnagar, Gujarat, India, 2007, p. 70.

Chapter 6.10

Gymnema sylvestre

Gymnema sylvestre (Retz.) R.Br. ex Sm. Asclepiadaceae

Common vernacular names: Bengali – Medhasingi; English – Periploca of the wood; Gujarati – Kaavalee, Medhasinge; Hindi – Gudmaar, Medhaa Singee; Kannada – Kadhasige; Malayalam – Cakkarakkolli, Madhunaashini; Marathi – Kaavalee, Medhaashingi; Tamil – Shirukurum Kaay, Shakkar-aikkolli; Telugu – Podapatro [1] (Fig. 6.10.1).

FIG. 6.10.1 *Gymnema sylvestre* leaf.

Macroscopic description: Shape – Elliptical or ovate; **Apex** – Acute or shortly acuminate; **Base** – Rounded or cordate; **Margin** – Entire; **Venation** – Reticulate; **Petiole** – Short; **Colour** – Dark green on upper side and light green on lower side, pubescent on both surfaces; **Odour** – Unpleasant.

Major chemical constituents: Triterpenoid saponins of gymnemic acid A, B, C and D with sugar residues such as glucuronic acid, galacturonic acid, ferulic and angelic acids attached as carboxylic acids [2].

Traditional uses: Antidiabetic [1].

Powdered Crude Drug Microscopy of Leaves and Barks. https://doi.org/10.1016/B978-0-12-818092-1.00034-8

TABLE 6.10.1 Specific as well as general characters of leaf powder of *Gymnema sylvestre.*

Features	
Epidermis	**Trichomes**
Thin walled	Abundant
Slightly sinous	Simple
Striated cuticle	Covering
	Thick walled
Stomata	Multicellular
Anomocytic	Unicellular
Actinocytic	Uniseriate
	Straight
Calcium oxalate crystals	Bent
Cluster	Curved
Rosette	Warty
	Striated walls
Lamina	Pedestial base
Spongy parenchyma	Pointed apex
	Buldging base

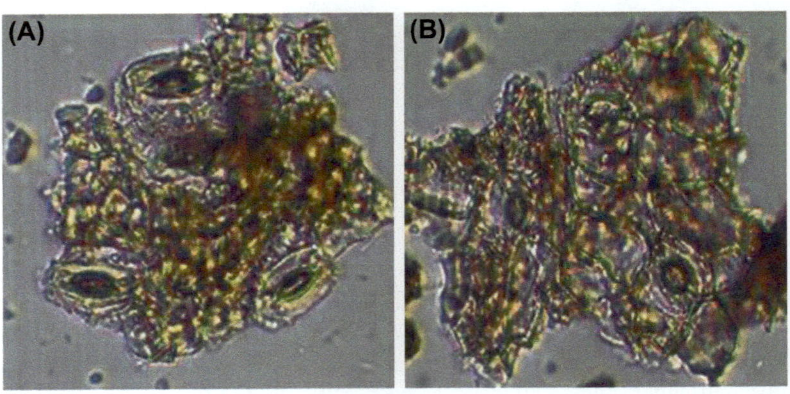

FIG. 6.10.2 (A–B) – Anomocytic and actinocytic stomata present in *Gymnema sylvestre* leaf powder.

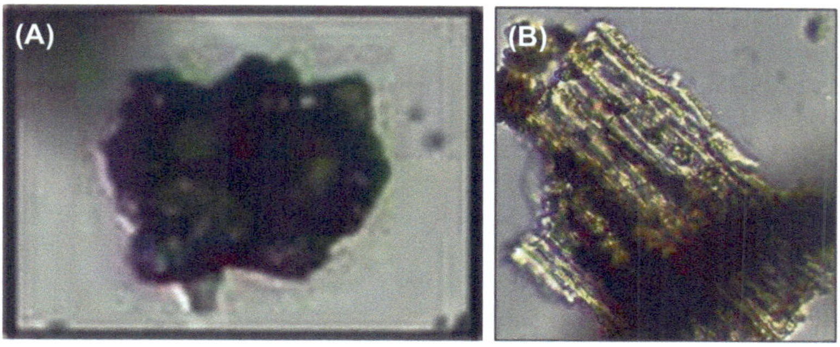

FIG. 6.10.3 (A–B) – Cluster crystal and rosette crystal present in *Gymnema sylvestre* leaf powder.

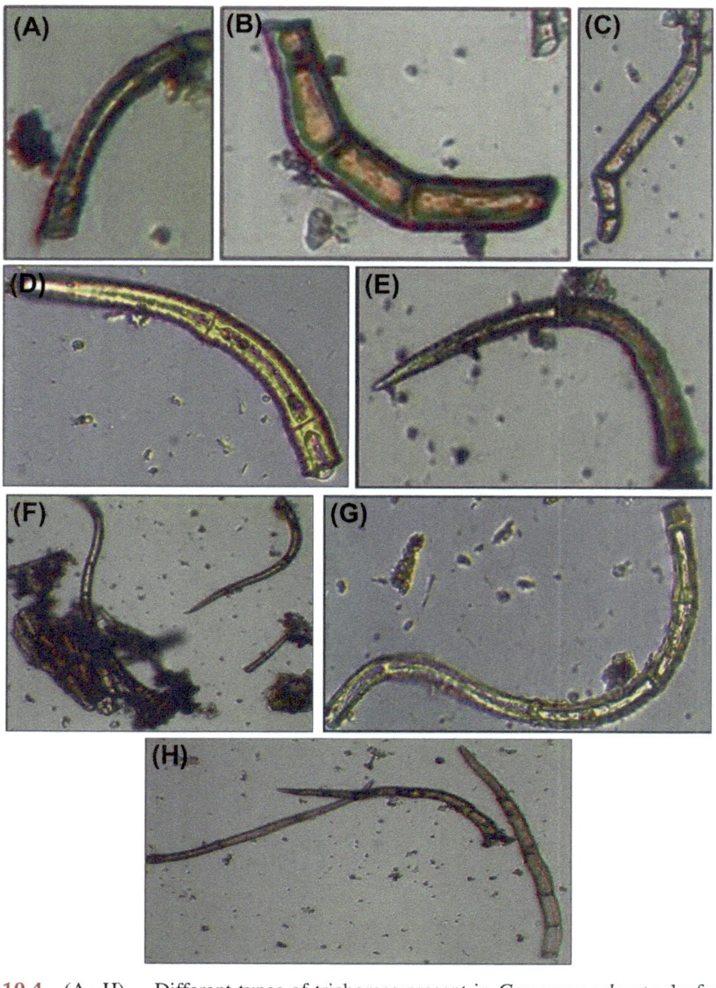

FIG. 6.10.4 (A–H) – Different types of trichomes present in *Gymnema sylvestre* leaf powder.

Powder microscopic description: Green colour powder showing fragments of upper and lower epidermis, thin walled, slightly sinous with striated cuticle embedded with anomocytic and actinocytic stomata; Plenty of simple, covering, thick walled multicellular, uniseriate, straight or irregularly bent, trichomes, occasionally exhibiting striated walls or pedestial base; unicellular trichomes are few, thick walled with pointed apex and bulging base; Transversely cut fragments of spongy parenchyma cells with rosette crystals of calcium oxalate and cluster, rosette and few prismatic crystals of calcium oxalate scattered as such throughout the powder [3]. The general as well as specific features of *G. sylvestre* leaf powder are illustrated in Table 6.10.1 and Figs. 6.10.2–6.10.4A–H respectively.

References

[1] Government of India, Department of AYUSH, Ayurvedic Pharmacopoeia of India. Part I, vol. V, Ministry of Health and Welfare, Ghaziabad, 1999–2017, p. 129.

[2] A.K. Gupta, N. Tandon, M. Sharma, Quality Standards of Indian Medicinal Plants, vol. 1, Medicinal Plant Unit, ICMR, New Delhi, India, 2008, p. 102.

[3] M.G. Chauhan, A.P.G. Pillai, Microscopic Profile of Powdered Drugs Used in Indian System of Medicine, Leaf Drugs, vol. 2, Institute of Ayurvedic Medicinal Plant Sciences, Jamnagar, Gujarat, India, 2007, p. 92.

Chapter 6.11

Jasminum officinale

Jasminum officinale L. Oleaceae

Common vernacular names: Assamese – Yasmeen; Bengali – Chamelee; English – Jasmine; Gujarati – Chamelee; Hindi – Chamelee; Kannada – Jati Maltiga, Sanna Jati Mallige; Malayalam – Pichi; Marathi – Chamelee; Punjabi – Chamelee; Tamil – Pichi, Jatimalli; Telugu – Jati, Sannajati [1] (Fig. 6.11.1).

Macroscopic description: Shape – Ovate to lanceolate; **Apex** – Acute to acuminate; **Base** – Symmetrical; **Margin** – Entire; **Venation** – Lateral veins three arising from the base; **Petiole** – Minutely petiolule; **Colour** – Dark green; **Odour** – Characteristic.

Major chemical constituents: Oleuropein, ligstroside, demethyloleuropein, (2″ R)- and (2″S)-2″-methoxyoleuropein, ursolic acid [2].

Traditional uses: Wound healing, analgesic, mouth ulcers and dermatopathy [2].

FIG. 6.11.1 *Jasminum officinale* leaf.

Powdered Crude Drug Microscopy of Leaves and Barks. https://doi.org/10.1016/B978-0-12-818092-1.00035-X

TABLE 6.11.1 Specific as well as general characters of leaf powder of *Jasminum officinale*.

Features	
Epidermis	**Trichomes**
Straight	Few
Thick	Simple
Anticlinical	Covering
Stomata	Unicellular
Actinocytic	**Calcium oxalate crystals**
Diacytic	Rosette

Powder microscopic description: Dark green colour powder showing fragments of upper and lower epidermis showing straight, thick, anticlinical walls embedded with actinocytic, diacytic stomata and oil globules; Fragments of simple, covering and unicellular trichomes and few rosette crystals of calcium oxalate [3]. The general as well as specific features of *J. officinale* leaf powder are illustrated in Table 6.11.1 and Figs. 6.11.2–6.11.4 respectively.

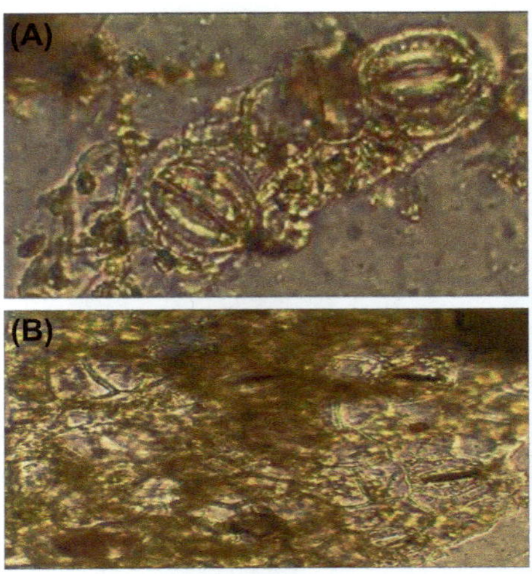

FIG. 6.11.2 (A–B) – Actinocytic and Diacytic stomata present in *Jasminum officinale* leaf powder.

FIG. 6.11.3 Covering trichome present in *Jasminum officinale* leaf powder.

FIG. 6.11.4 Rosette crystal present in *Jasminum officinale* leaf powder.

References

[1] Government of India, Department of AYUSH, Ayurvedic Pharmacopoeia of India. Part I, vol. III, Ministry of Health and Welfare, Ghaziabad, 1999–2017, p. 72.

[2] A.K. Gupta, N. Tandon, M. Sharma, Quality Standards of Indian Medicinal Plants, vol. 7, Medicinal Plant Unit, ICMR, New Delhi, India, 2008, p. 152.

[3] M.G. Chauhan, A.P.G. Pillai, Microscopic Profile of Powdered Drugs Used in Indian System of Medicine, Leaf Drugs, vol. 2, Institute of Ayurvedic Medicinal Plant Sciences, Jamnagar, Gujarat, India, 2007, p. 58.

Chapter 6.12

Murraya koenigii

Murraya koenigii (L.) Spreng. Rutaceae

Common vernacular names: Assamese – Narasingha; Bengali - Bansang, Kariaphulli; English - Curry leaf; Gujrati - Gornimb, Kadhilimdo; Hindi – Mitha neem, Kadhi Patta, Kadi Patta; Kannada – Karibaevu; Malyalam – Kariveppu; Marathi - Kadhinim, Poospala, Godnimb; Oriya – Bhursunga; Punjabi - Kadhi Patta; Tamil - Karivempu, Karuveppilei; Telugu - Karivepaku, Karivemu [1].

Macroscopic description: Shape – Ovate; **Apex** – Acute; **Base** – Asymmetrical; **Margin** – Entire; **Venation** – Reticulate; **Petiole** – Shortly petiolulate; **Colour** – Dark green when fresh; **Odour** – Characteristically aromatic.

Major chemical constituents: Alkaloids like koenidine, koenigine, koenimbine, mahanimbine, muconine murrayacine and volatile oils [2].

Traditional uses: Piles, Diarrhoea, Emesis [1] (Fig. 6.12.1).

FIG. 6.12.1 *Murraya koenigii* leaf.

Powdered Crude Drug Microscopy of Leaves and Barks. https://doi.org/10.1016/B978-0-12-818092-1.00036-1

Powder microscopic description: Dark green colour powder showing plenty of simple, unicellular, thick walled, sickle shaped or straight, trichomes with pointed apex; Prismatic crystals of calcium oxalate and fragments of epidermis with straight, anticlinical, slightly sinous walls embedded anomocytic stomata and oil globules [3]. The general as well as specific features of *M.koenigii* leaf powder are illustrated in Table 6.12.1 and Fig. 6.12.2A–C, 6.12.3 and 6.12.4A–B respectively.

TABLE 6.12.1 Specific as well as general characters of leaf powder of *Murraya koenigii.*

Features	
Epidermis	**Trichomes**
Straight	Simple
Slightly sinous	Abundant
Thick	Unicellular
Anticlinical walls	Covering
Stomata	Thick walled
Anomocytic	Sickle shaped
Calcium oxalate crystals	Straight
Prismatic	Pointed apex

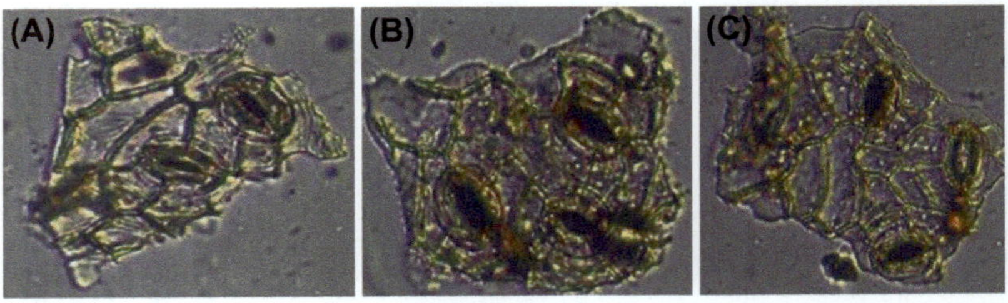

FIG. 6.12.2 (A–C) – Anomocytic stomata present in *Murraya koenigii* leaf powder.

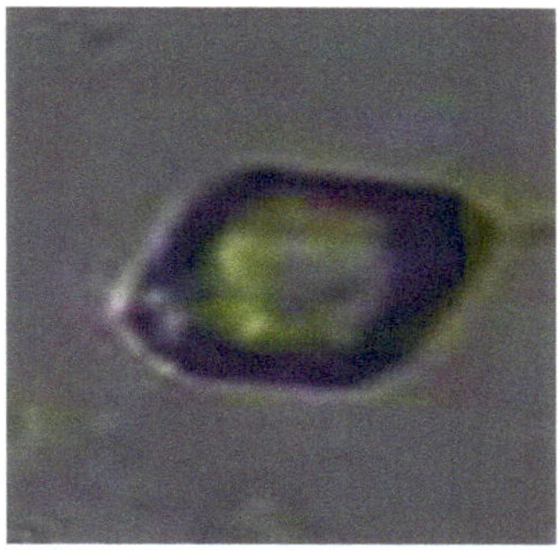

FIG. 6.12.3 Prismatic crystal present in *Murraya koenigii* leaf powder.

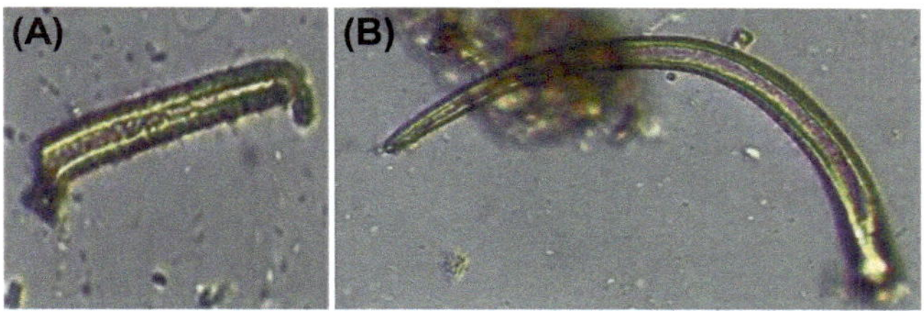

FIG. 6.12.4 (A–B) – Covering trichomes present in *Murraya koenigii* leaf powder.

References

[1] Government of India, Department of AYUSH, Ayurvedic Pharmacopoeia of India. Part I, vol. VI, Ministry of Health and welfare, Ghaziabad, 1999-2017, p. CLXXX.

[2] A.K. Gupta, N. Tandon, M. Sharma, Quality Standards of Indian Medicinal Plants, vol. 1, Medicinal plant unit, ICMR, New Delhi, India, 2008, p. 142.

[3] M.G. Chauhan, A.P.G. Pillai, Microscopic Profile of Powdered Drugs Used in Indian System of Medicine, Leaf Drugs, vol. 2, Institute of Ayurvedic Medicinal Plant Sciences, Jamnagar, Gujarat, India, 2007, p. 66.

Chapter 6.13

Piper betle

Piper betle L. Piperaceae

Common vernacular names: Assamese – Pan; Bengali – Pan; English – Betel Leaf; Gujrati – Pan; Hindi: Pan; Kannada – Veelyadele Ele; Malyalam – Vettila; Marathi – Pan, Nagvel, Vidyachepan; Punjabi – Pan; Tamil – Vettilai; Telugu – Tamalapaku, Tamulapaku [1].

Macroscopic description: Shape – Cordate to obliquely ovate; **Apex** – Acute to acuminate; **Base** – Undulate; **Margin** – Entire; **Venation** – Reticulate, multicostate; **Petiole** – Cylindrical; **Colour** – Upper surface yellowish green, lower lighter in colour; **Odour** – Pleasant and aromatic (Fig. 6.13.1).

Major chemical constituents: Eugenol, eugenyl acetate, β-sitosterol [2].

Traditional uses: Antifertility, Antimicrobial and Filariasis [2].

Powder microscopic description: Yellowish green colour powder showing fragments of upper and lower epidermis with straight, thick anticlinical walls embedded with oil globules, brownish content and actinocytic

FIG. 6.13.1 *Piper betle* leaf.

Powdered Crude Drug Microscopy of Leaves and Barks. https://doi.org/10.1016/B978-0-12-818092-1.00037-3

TABLE 6.13.1 Specific as well as general characters of leaf powder of *Piper betle*.

Features		
Epidermis	**Trichomes**	**Cell contents**
Straight	Abundant	Volatile oil
Thick	Simple	Brownish matter
Anticlinical walls	Covering	Mucilage
Stomata	Unicellular	
Actinocytic	Multicellular	
	Irregularly bent	**Xylem vessels**
Calcium oxalate crystals	Curved	Spiral
Prismatic	Pointed apex	

stomata; Plenty of unicellular to multicellular, simple covering trichomes; Prismatic crystals of calcium oxalate along with mucilage and spherical oil cells; Longitudinally cut fragments of spiral xylem vessels [3]. The general as well as specific features of *P.betle* leaf powder are illustrated in Table 6.13.1 and Fig. 6.13.2, 6.13.3, 6.13.4A-C, 6.13.5 and 6.13.6 respectively.

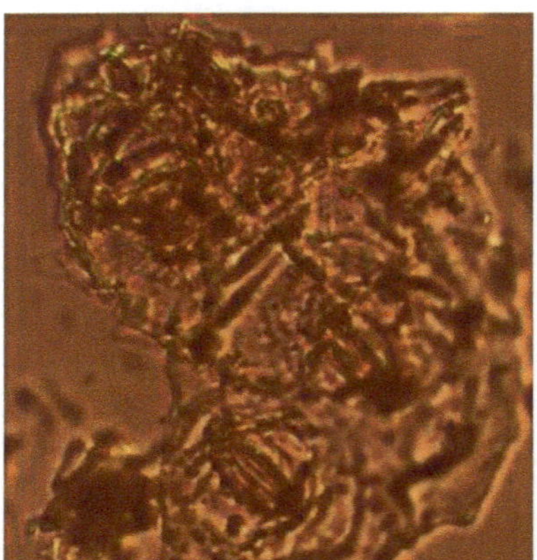

FIG. 6.13.2 Actinocytic stomata present in *Piper betle* leaf powder.

FIG. 6.13.3 Prismatic crystal present in *Piper betle* leaf powder.

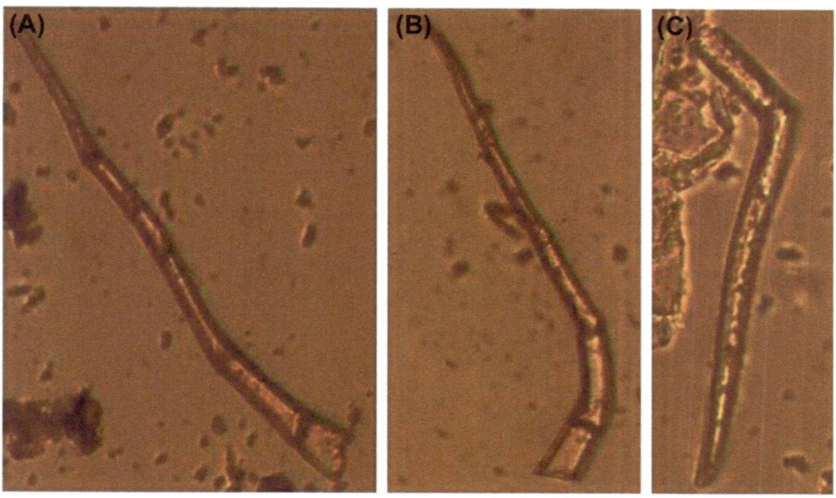

FIG. 6.13.4 (A–C) – Covering trichomes present in *Piper betle* leaf powder.

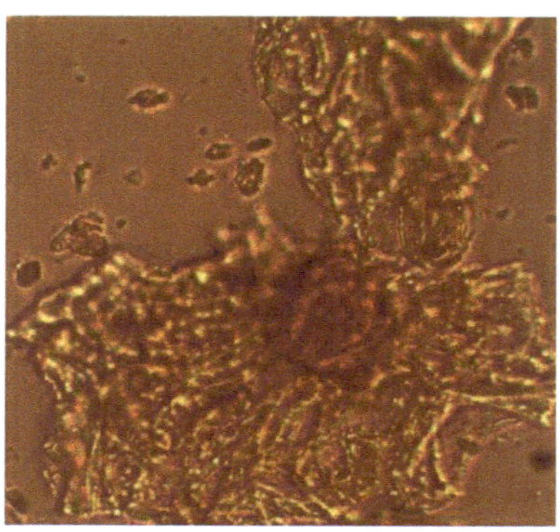

FIG. 6.13.5 Brownish matter in epidermis present in *Piper betle* leaf powder.

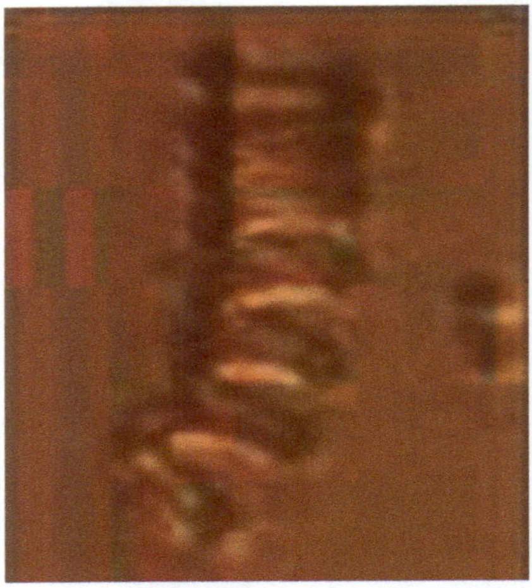

FIG. 6.13.6 Spiral xylem vessel present in *Piper betle* leaf powder.

References

[1] Government of India, Department of AYUSH, Ayurvedic Pharmacopoeia of India. Part I, vol. III, Ministry of Health and welfare, Ghaziabad, 1999-2017, p. 133.

[2] A.K. Gupta, N. Tandon, M. Sharma, Quality Standards of Indian Medicinal Plants, vol. 7, Medicinal plant unit, ICMR, New Delhi, India, 2008, p. 198.

[3] M.G. Chauhan, A.P.G. Pillai, Microscopic Profile of Powdered Drugs Used in Indian System of Medicine Leaf Drugs, vol. 2, Institute of Ayurvedic Medicinal Plant Sciences, Jamnagar, Gujarat, India, 2007, p. 162.

Chapter 6.14

Plantago lanceolata

Plantago lanceolata L. Plantaginaceae

Common vernacular names: Bengali – Bartung; English – Ribwort; Hindi – Baltanga, Jangali isabgola; Kannada – Siriportlagida; Marathi – Baltang; Punjabi – Kashur-gul; Telugu – Adavi ishapugorulu [1].

Macroscopic description: Shape – Rosette of lanceolate ribbed leaves; **Apex** – Truncate; **Base** – Lamina tapering downwards in a short broad and curved stalk; **Margin** – Entire; **Venation** – Convergent; **Petiole** – Margined; **Colour** – Green; **Odour** – Characteristic.

Major chemical constituents: Chlorogenic acid, chrysophanic acid, emodin, luteolin, plantaginin, scutellarin, aesculetin [1].

Traditional uses: Piles, Otalgia, Menorrhagia [1] (Fig. 6.14.1).

FIG. 6.14.1 *Plantago lanceolata* leaf.

Powdered Crude Drug Microscopy of Leaves and Barks. https://doi.org/10.1016/B978-0-12-818092-1.00038-5

TABLE 6.14.1 Specific as well as general characters of leaf powder of *Plantago lanceolata*.

Features	
Epidermis	**Stomata**
Wavy	Paracytic
Thin walled	
Anticlinical	**Cell contents**
Calcium oxalate crystals	Brownish matter
Prismatic	Mucilage

Powder microscopic description: Green colour powder showing fragments of upper and lower epidermis with wavy walls embedded with paracytic stomata; Very large prismatic crystals of calcium oxalate; Brownish matter, mucilage cells and devoid of trichomes. The general as well as specific features of *P.lanceolata* leaf powder are illustrated in Table 6.14.1 and Fig. 6.14.2, 6.14.3 and 6.14.4 respectively.

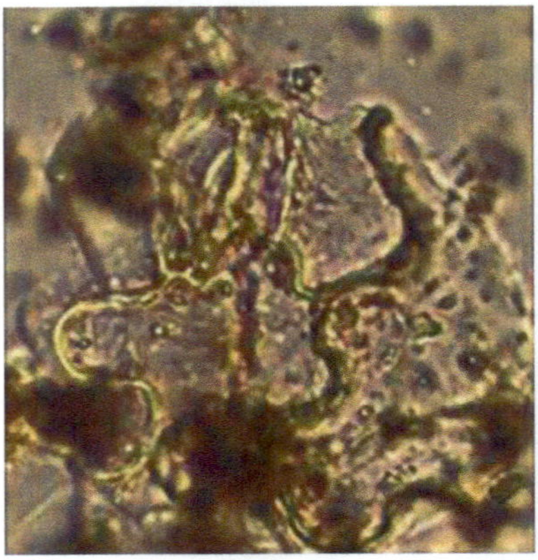

FIG. 6.14.2 Paracytic stomata present in *Plantago lanceolata* leaf powder.

FIG. 6.14.3 Prismatic crystal present in *Plantago lanceolata* leaf powder.

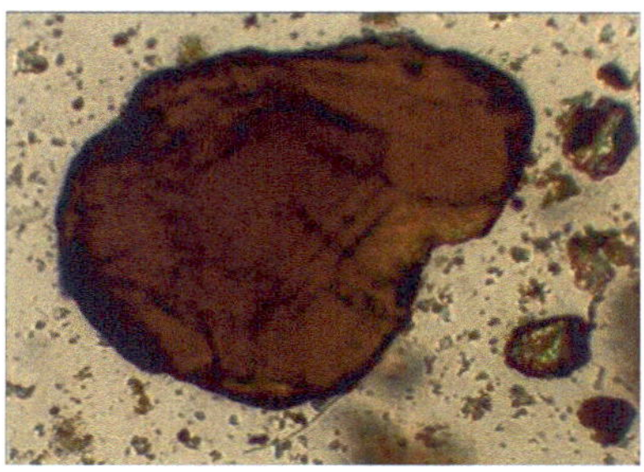

FIG. 6.14.4 Brownish matter present in *Plantago lanceolata* leaf powder.

Reference

[1] Government of India, Department of AYUSH, Ayurvedic Pharmacopoeia of India. Part I, vol. VI, Ministry of Health and welfare, Ghaziabad, 1999-2017, p. CLXXIII.

Chapter 6.15

Pongamia pinnata

Pongamia pinnata (L.) Pierre Fabaceae

Common vernacular names: Assamese – Korach; Bengali – Dahara karanja, Karanja, Natakaranja; English – Smooth leaved pongamia; Gujrati – Kanaji, Kanajo; Hindi – Karuaini, Dithouri; Kannada – Honge, Hulagilu; Malayalam – Pungu, Ungu, Unu, Avittal; Marathi – Karanja; Oriya – Karanja; Punjabi – Karanj; Tamil – Pungai, Pongana; Telugu – Ganuga, Kanugu [1].

Macroscopic description: Shape – Ovate to elliptic; **Apex** – Acute; **Base** – Rounded to oblique; **Margin** – Smooth**; Venation** – Alternate; **Petiole** – Short; **Colour** – Dark green; **Odour** – Characteristic.

Major chemical constituents: Furanoflavone -3' - methoxy pongapin in addition to Karanjin, Kanjone and its two isomers 7-Methoxyfurano-(4″,5″,6,5) - flavone and 8-Methoxyfurano-(4″, 5″, - 6, 5)-flavone and 8 - methoxyfurano- (4″, 5" - 6, 7) -flavone [1].

Traditional uses: Leprosy, rheumatic pain [1] (Fig. 6.15.1).

FIG. 6.15.1 *Pongamia pinnata* leaf.

Powdered Crude Drug Microscopy of Leaves and Barks. https://doi.org/10.1016/B978-0-12-818092-1.00039-7

Powder microscopic description: Dark green colour powder showing fragments of upper and lower epidermis with wavy walls embedded with paracytic stomata; Simple, unicellular, covering trichomes with pointed apex; Prismatic and rod shaped crystals of calcium oxalate and presence of xylem vessels with spiral thickenings. The general as well as specific features of *P.pinnata* leaf powder are illustrated in Table 6.15.1 and Figs. 6.15.2–6.15.6 respectively.

TABLE 6.15.1 Specific as well as general characters of leaf powder of *Pongamia pinnata*.

Features		
Epidermis	**Trichomes**	**Calcium oxalate crystals**
Thick	Few	Prismatic
Wavy	Simple	Rod shaped
	Unicellular	
Stomata	Uniseriate	**Xylem vessels**
Paracytic	Covering	Spiral
	Slightly bent	
	Pointed apex	

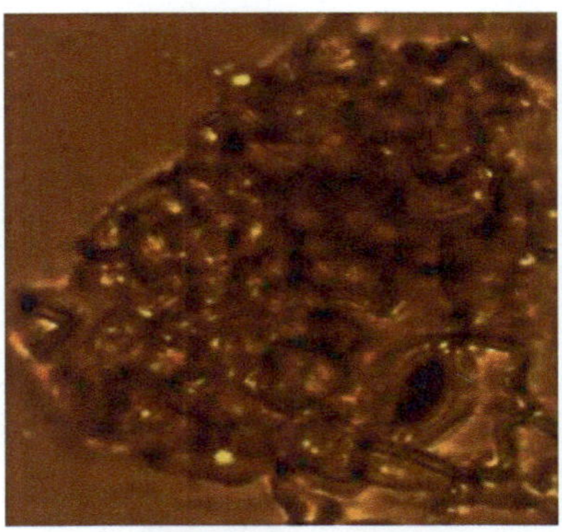

FIG. 6.15.2 Paracytic stomata present in *Pongamia pinnata* leaf powder.

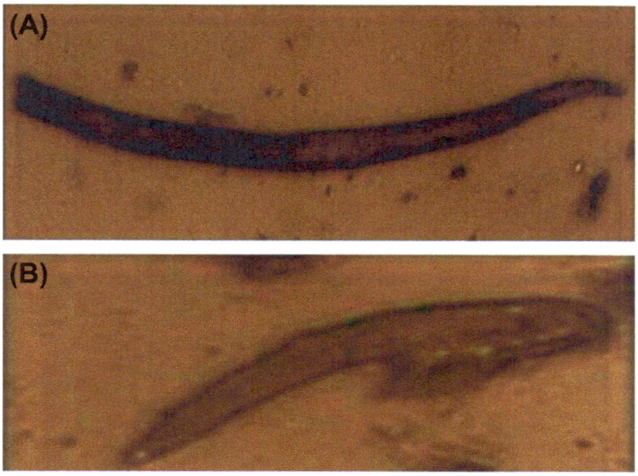

FIG. 6.15.3 (A–B) – Covering trichomes present in *Pongamia pinnata* leaf powder.

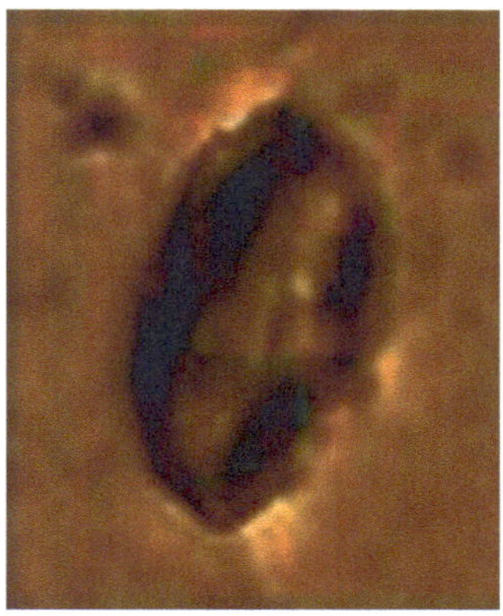

FIG. 6.15.4 Prismatic crystals present in *Pongamia pinnata* leaf powder.

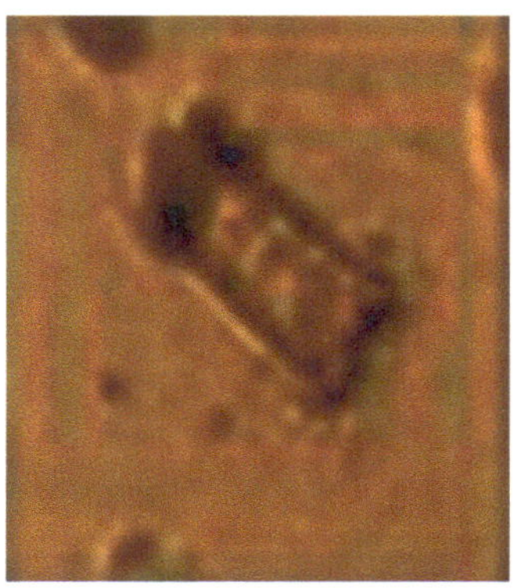

FIG. 6.15.5 Acicular crystals present in *Pongamia pinnata* leaf powder.

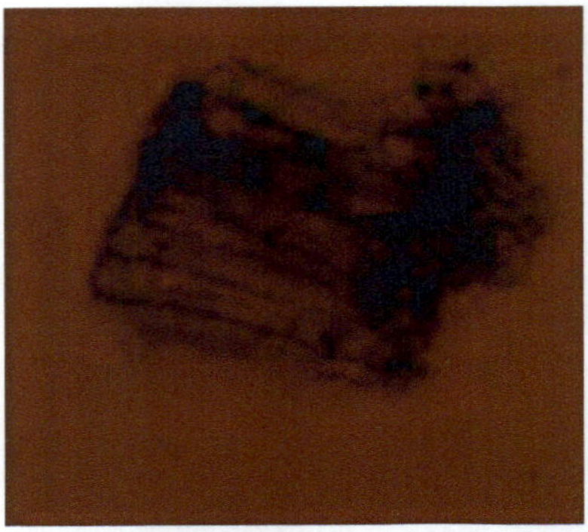

FIG. 6.15.6 Spiral xylem vessels present in *Pongamia pinnata* leaf powder.

Reference

[1] Government of India, Department of AYUSH, Ayurvedic Pharmacopoeia of India. Part I, vol. II, Ministry of Health and welfare, Ghaziabad, 1999–2017, p. 87.

Chapter 6.16

Prosopis cineraria

Prosopis cineraria (L.) Druce Leguminosae

Common vernacular names: Bengali – Sain, Shami; English – Spunge tree; Gujrati – Kheejado, Sami; Hindi – Chhonkar, Sami, Chhikur, Jhand, Khejra; Kannada – Banni, Kabanni; Malyalam – Parampu, Tambu, Vahni; Marathi – Sami, Saunder; Oriya – Shami; Punjabi – Jand; Tamil – Vanni; Telugu – Jammi [1].

Macroscopic description: Shape – Oblong and rounded; **Apex** – Mucronate; **Base** – Rounded; **Margin** – Entire; **Petiole** – Almost triangular; **Colour** – Dark green; **Odour** – Not characteristic (Fig. 6.16.1).

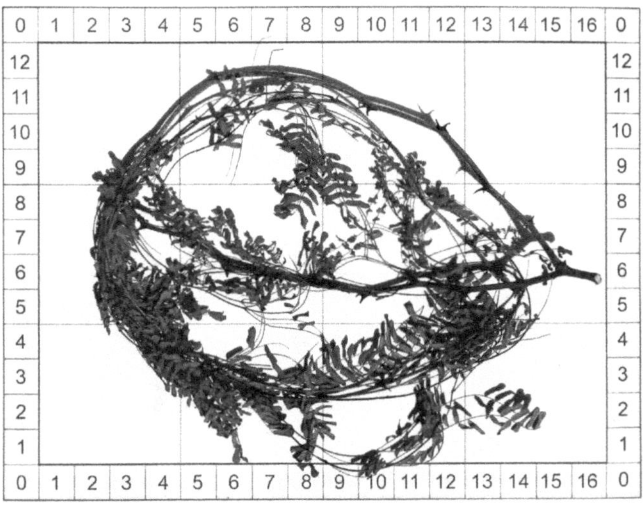

FIG. 6.16.1 *Prosopis cineraria* leaf.

Major chemical constituents: Tannin rich foliage, volatile fatty acid [1].

Traditional uses: Piles, Diarrhoea, vertigo, worm infestation [1].

Powder microscopic description: Dark green colour powder showing fragments of upper and lower epidermis with wavy walls embedded with

Powdered Crude Drug Microscopy of Leaves and Barks. https://doi.org/10.1016/B978-0-12-818092-1.00040-3

paracytic stomata; Few, simple, covering, thick walled trichomes; Rosette crystals of calcium oxalate; Longitudinally cut fragments of spiral xylem vessels and few, oval to circular, simple and compound starch grains. The general as well as specific features of *P.cineraria* leaf powder are illustrated in Table 6.16.1 and Figs. 6.16.2A–B, 6.16.3, 6.16.4, 6.16.5 and 6.16.6 respectively.

TABLE 6.16.1 Specific as well as general characters of leaf powder of *Prosopis cineraria*.

Features		
Epidermis	**Trichomes**	**Starch grains**
Wavy	Few	Abundant
Anticlinical walls	Simple	Simple
Stomata	Unicellular	Spherical
Paracytic	Covering	Compound
	Thick walled	Oval
Calcium oxalate crystals		**Xylem vessels**
Rosette		Spiral

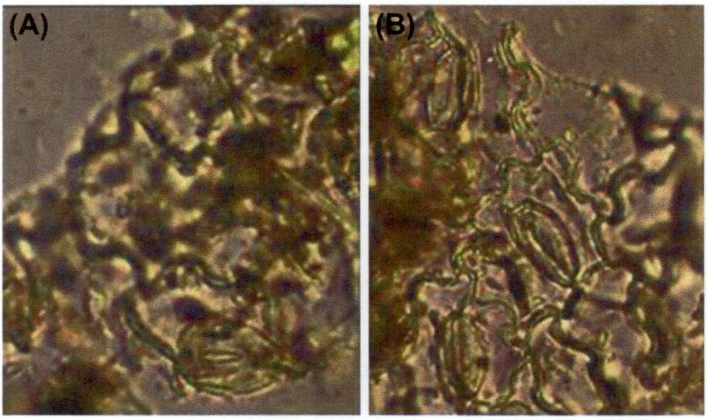

FIG. 6.16.2 (A–B) – Paracytic stomata present in *Prosopis cineraria* leaf powder.

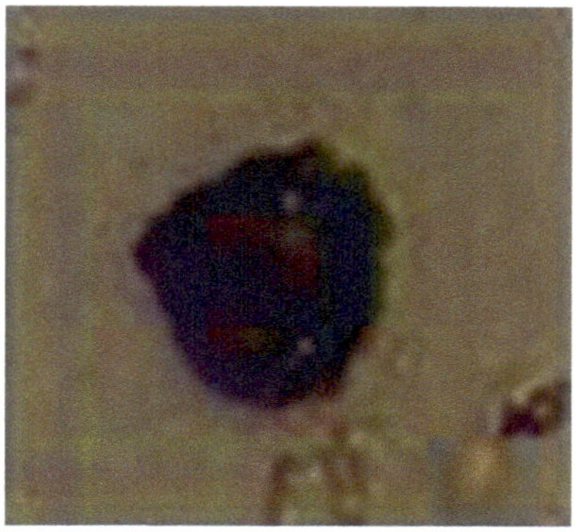

FIG. 6.16.3 Rosette crystals present in *Prosopis cineraria* leaf powder.

FIG. 6.16.4 Covering trichome present in *Prosopis cineraria* leaf powder.

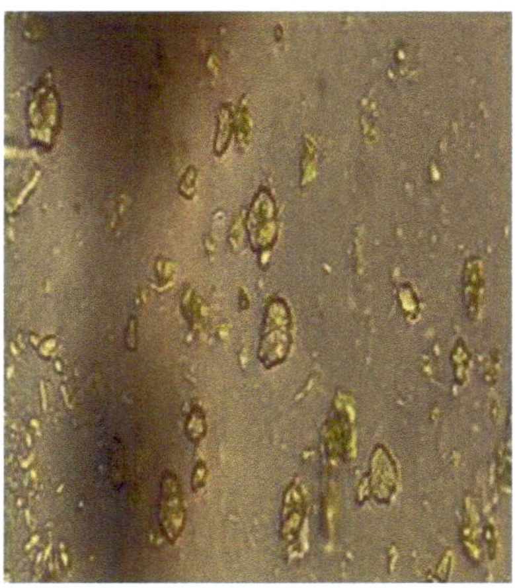

FIG. 6.16.5 Starch grains present in *Prosopis cineraria* leaf powder.

FIG. 6.16.6 Spiral xylem vessels present in *Prosopis cineraria* leaf powder.

Reference

[1] Government of India, Department of AYUSH, Ayurvedic Pharmacopoeia of India. Part I, vol. VI, Ministry of Health and welfare, Ghaziabad, 1999-2017, p. CLXXVII.

Chapter 6.17

Punica granatum

Punica granatum L. Punicaceae

Common vernacular names: Assamese – Dalim; Bengali – Dadima, Dalim, Dalimgach; English – Pomenagrate; Gujrati –Dadam, Dadam phala; Hindi – Anar, Anar-ke-per; Kannada: Dalimba, Dalimbe haonu; Malayalam – Mathalam; Marathi – Dalimba; Oriya – Dalimba; Punjabi – Anar; Tamil – Madulam Pazham; Telugu – Dadimbakaya, Dadimma [1].

Macroscopic description: Shape – Oblong to lanceolate; **Apex** – Acute to obtuse; **Base** – Cuneate; **Margin** – Entire; **Venation** – Pinnate; **Petiole** – Short; **Colour** – Greyish green to yellowish green; **Odour** – Characteristic.

Major chemical constituents: β-sitosterol and tannins [1]

Traditional uses: Stomach disorders or Diarrhoea [1] (Fig. 6.17.1).

FIG. 6.17.1 *Punica granatum* leaf.

Powdered Crude Drug Microscopy of Leaves and Barks. https://doi.org/10.1016/B978-0-12-818092-1.00041-5

TABLE 6.17.1 Specific as well as general characters of leaf powder of *Punica granatum*.

Features	
Epidermis	**Stomata**
Wavy	Anisocytic
Beaded	Anomocytic
Calcium oxalate crystal	
Prismatic	
Rosette	

Powder microscopic description: Greyish green colour powder showing fragments of upper and lower epidermis with beaded wavy walls embedded with anisocytic and anomocytic stomata; Prismatic and rosette crystals of calcium oxalate; Few simple starch grains and xylem vessels and devoid of trichomes [2]. The general as well as specific features of *P.granatum* leaf powder are illustrated in Table 6.17.1 and Figs. 6.17.2, 6.17.3 and 6.17.4 respectively.

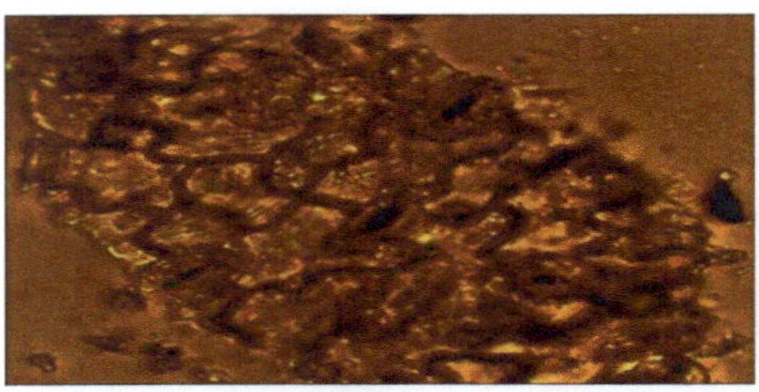

FIG. 6.17.2 Anomocytic stomata present in *Punica granatum* leaf powder.

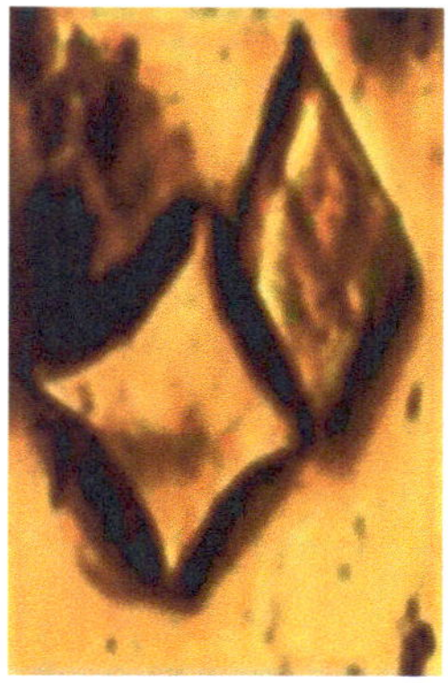

FIG. 6.17.3 *Prismatic* crystal present in *Punica granatum* leaf powder.

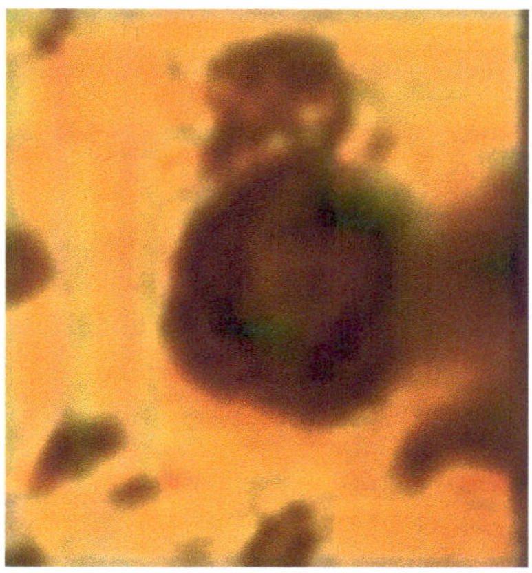

FIG. 6.17.4 *Rosette* crystal present in *Punica granatum* leaf powder.

References

[1] Government of India, Department of AYUSH, Ayurvedic Pharmacopoeia of India. Part I, vol. IV, Ministry of Health and welfare, Ghaziabad, 1999-2017, p. 25.

[2] M.G. Chauhan, A.P.G. Pillai, Microscopic Profile of Powdered Drugs Used in Indian System of Medicine, Leaf Drugs, vol. 2, Institute of Ayurvedic Medicinal Plant Sciences, Jamnagar, Gujarat, India, 2007, p. 38.

Chapter 6.18

Ricinus communis

Ricinus communis L. Euphorbiaceae

Common vernacular names: Assamese – Erri; Bengali – Bherenda; English – Castor Oil Plant; Gujrati – Erando; Hindi – Erand, Rendee, Andu; Kannada – Harlu; Malayalam – Ambanakka, Avanakku; Marathi - Erand, Erandee; Oriya – Bheranda; Punjabi – Erand; Tamil – Amanakku; Telugu – Amudanu, Amudmuchetu [1].

Macroscopic description: Shape – Palmately lobed; **Apex** – Acute to acuminate; **Base** – Cordate; **Margin** – Serrate; **Venation** – Palmately veined; **Petiole** – Very long, cylindrical, faintly longitudinally striated; **Colour** – Upper surface dark green and lower paler; **Odour** – Not characteristic (Fig. 6.18.1).

Major chemical constituents: Ricinine and rutin [2].

Traditional uses: Lower abdominal colic, dysuria, anthelmentic [2].

Powder microscopic description: Green colour powder showing fragments of upper and lower epidermis, slightly sinous walled exhibiting paracytic stomata; Simple, unicellular, covering trichomes with pointed apex; Needle shaped crystals of calcium oxalate and presence of annular xylem vessels. The general as well as specific features of *R.communis* leaf powder are illustrated in Table 6.18.1 and Figs. 6.18.2, 6.18.3 and 6.18.4 respectively.

Powdered Crude Drug Microscopy of Leaves and Barks. https://doi.org/10.1016/B978-0-12-818092-1.00042-7

FIG. 6.18.1 *Ricinus communis* leaf.

TABLE 6.18.1 Specific as well as general characters of leaf powder of *Ricinus communis*.

Features		
Epidermis	**Trichomes**	**Calcium oxalate crystals**
Slightly sinous walled	Simple	Needle
	Covering	
Stomata	Uniseriate	**Xylem vessels**
Paracytic	Unicellular	Annular
	Pointed apex	
	Curved	

FIG. 6.18.2 Paracytic stomata present in *Ricinus communis* leaf powder.

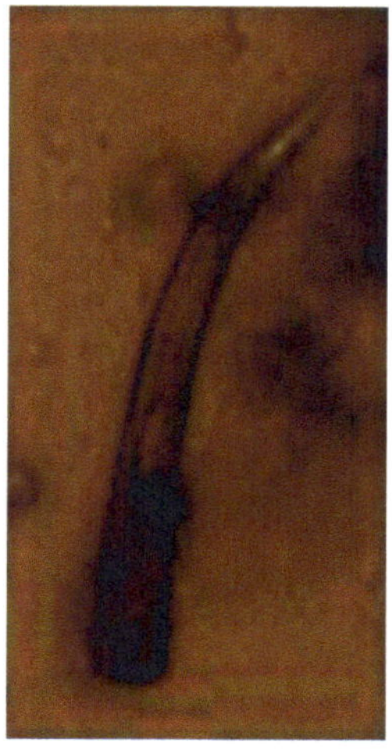

FIG. 6.18.3 Covering trichome present in *Ricinus communis* leaf powder.

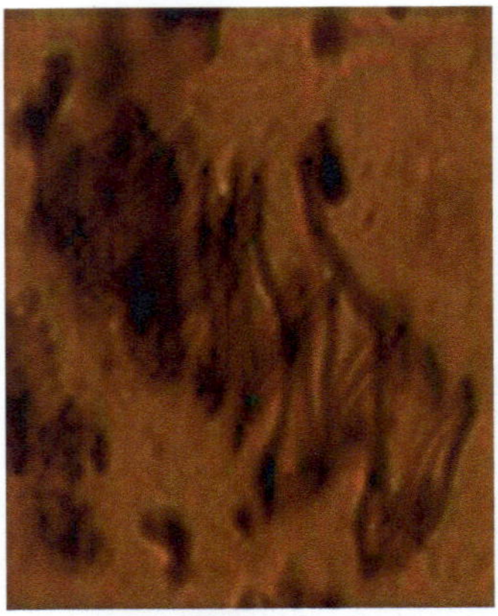

FIG. 6.18.4 Annular xylem vessel present in *Ricinus communis* leaf powder.

References

[1] Government of India, Department of AYUSH, Ayurvedic Pharmacopoeia of India. Part I, vol. III, Ministry of Health and welfare, Ghaziabad, 1999-2017, p. 49.

[2] A.K. Gupta, N. Tandon, M. Sharma, Quality Standards of Indian Medicinal Plants, vol. 10, Medicinal plant unit, ICMR, New Delhi, India, 2008, p. 303.

Chapter 6.19

Sesbania sesban

Sesbania sesban (L.) Merr Fabaceae

Common vernacular names: Bengali – Jayanti; Gujrati – Rajashinganee, Jayanti; Hindi – Jaita, jayata; Kannada – Arinintajinamgi, Karijimangai, Arishimajingai; Malayalam – Semp, Atti, Itthikkanni; Marathi – Jait; Oriya – Jayantipatra; Punjabi – Jainta; Tamil – Karum-sempai; Telugu – Sominta, Jalugu, Nelichettu [1].

Macroscopic description: Shape – Linear oblong; **Apex** – Obtuse; **Base** – Rounded; **Margin** – Entire; **Venation** – Lateral veins obscure; **Petiole** – Minutely petiolule; **Colour** – Faint green; **Odour** – No odour.

Major chemical constituents: Chikusetsusaponin IVa [2].

Traditional uses: Thyroid disorders, dysuria [2] (Fig. 6.19.1).

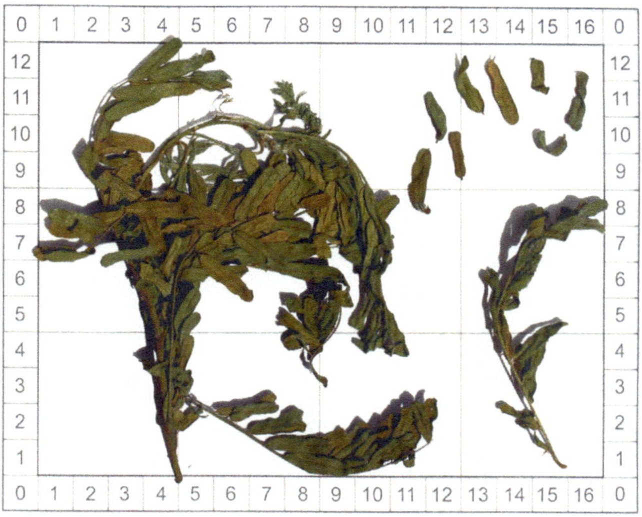

FIG. 6.19.1 *Sesbania sesban* leaf.

Powder microscopic description: Green colour powder showing fragments of upper and lower epidermis with straight to slightly sinous anticlinical walls

Powdered Crude Drug Microscopy of Leaves and Barks. https://doi.org/10.1016/B978-0-12-818092-1.00043-9

embedded with anomocytic and anisocytic stomata and parenchymatous cells filled with idioblast containing tannin cells; Simple, covering, unicellular, straight or slightly bent, warty trichomes; Prismatic and cluster crystals of calcium oxalate and longitudinally cut fragments of spiral and annular xylem vessels. The general as well as specific features of *S.sesban* leaf powder are illustrated in Table 6.19.1 and Figs. 6.19.2A–B, 6.19.3, 6.19.4, 6.19.5, 6.19.6 and 6.19.7A–E respectively.

TABLE 6.19.1 **Specific as well as general characters of leaf powder of *Sesbania sesban*.**

Epidermis	**Trichomes**	**Calcium oxalate crystals**
Straight	Abundant	Prismatic
Slightly sinous	Simple	Cluster
Anticlinical	Covering	
	Unicellular	**Cell contents**
Stomata	Straight	Tannin cell
Anisocytic	Bent	
Anomocytic	Warty	**Xylem vessels**
	Thick walled	Spiral
	Irregularly bent	Annular
	Pointed apex	
	Spear shaped	

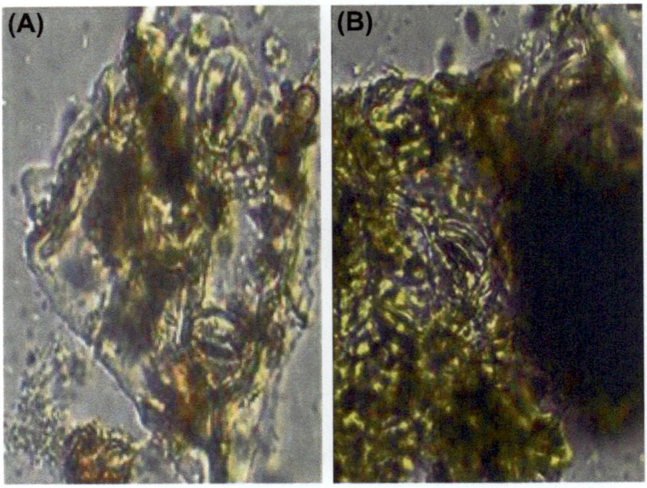

FIG. 6.19.2 (A–B) – Stomata present in *Sesbania sesban* leaf powder.

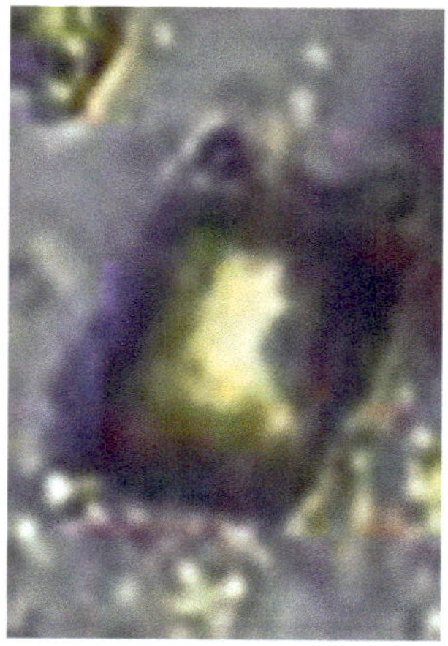

FIG. 6.19.3 Prismatic crystal present in *Sesbania sesban* leaf powder.

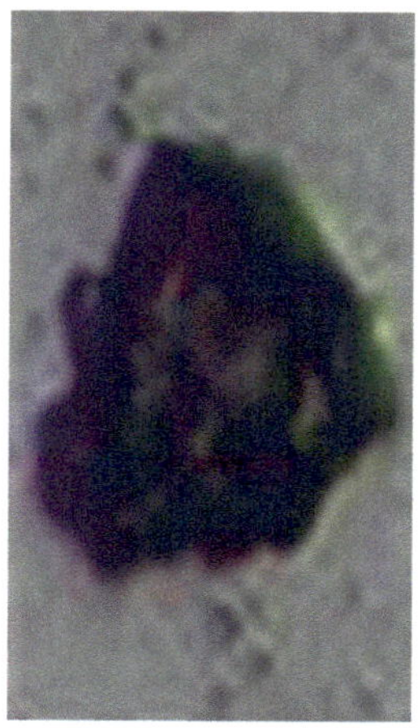

FIG. 6.19.4 Cluster crystal present in *Sesbania sesban* leaf powder.

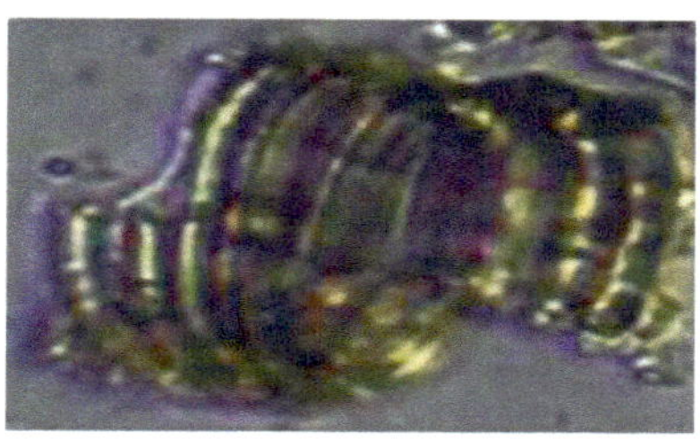

FIG. 6.19.5 Spiral xylem vessel present in *Sesbania sesban* leaf powder.

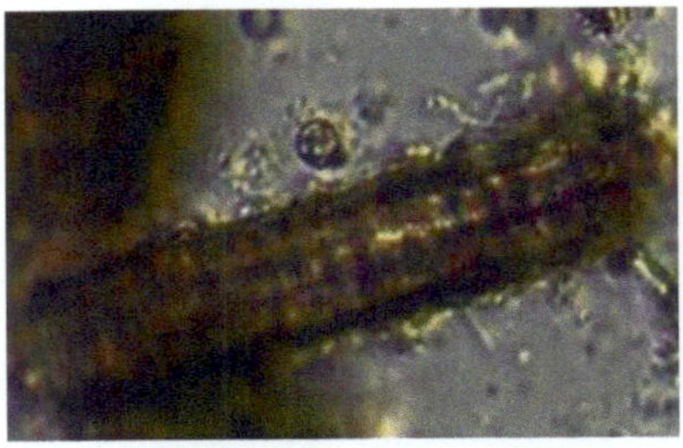

FIG. 6.19.6 Annular xylem vessels present in *Sesbania sesban* leaf powder.

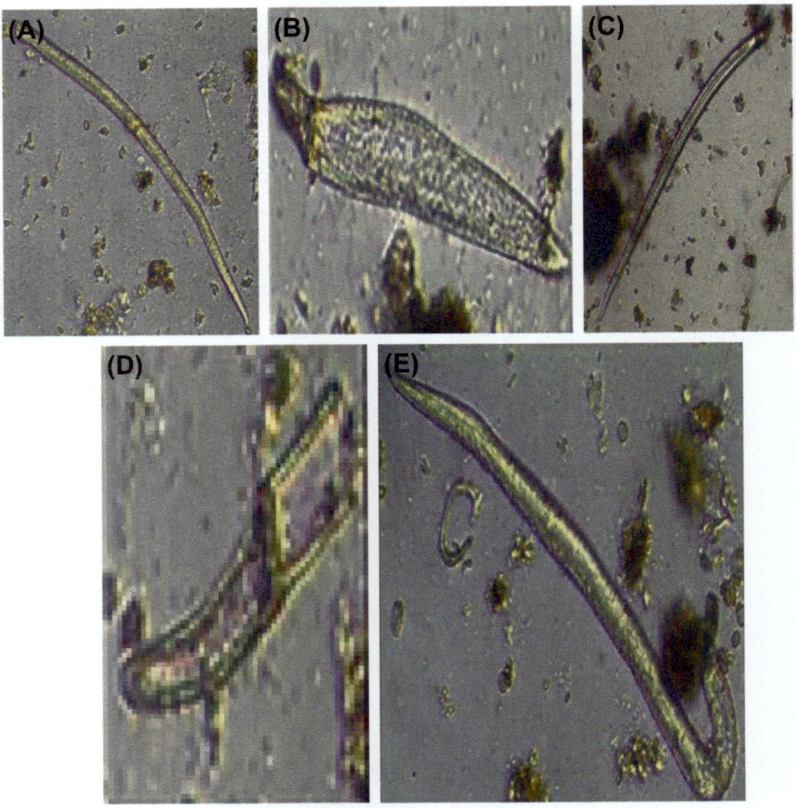

FIG. 6.19.7 Covering trichomes present in *Sesbania sesban* leaf powder.

References

[1] Government of India, Department of AYUSH, Ayurvedic Pharmacopoeia of India. Part I, vol. II, Ministry of Health and welfare, Ghaziabad, 1999-2017, p. 64.

[2] A.K. Gupta, N. Tandon, M. Sharma, Quality Standards of Indian Medicinal Plants, vol. 6, Medicinal plant unit, ICMR, New Delhi, India, 2008, p. 269.

Chapter 6.20

Sphaeranthus indicus

Sphaeranthus indicus L. Asteraceae

Common vernacular names: Assamese – Kamadarus; Bengali – Surmuriya, Chhagal Nadi, Mudmudiya; Gujrati – Gorakhmundi; Hindi – Mundi; Kannada – Mundi; Malayalam - Mirnagnee, Atookamanni, Mirangnee; Marathi – Mundi, Baras Bondi; Oriya – Buikadam; Punjabi – Gorakhmundi; Tamil – Kotook, Karandai, Kottakarthai; Telugu – Bodasaramu, Bodataramu [1] (Fig. 6.20.1).

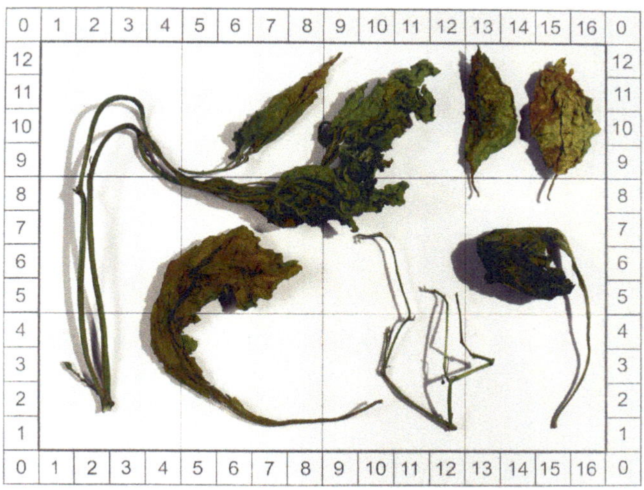

FIG. 6.20.1 *Sphaeranthus indicus* leaf.

Macroscopic description: Shape – Obovate-oblong; **Apex** – Rounded or subacute; **Base** – Narrowed to base; **Margin** – Dentate to serrate; **Petiole** – Cylindrical and long; **Colour** – Greenish brown; **Odour** – Slightly aromatic.

Traditional uses: Diarrhoea, bronchitis, piles [1].

Powder microscopic description: Greenish brown colour powder showing fragments of upper and lower epidermis with wavy walls embedded with minute sunken stomata and brownish matter; Prismatic crystals of calcium oxalate; Plenty of covering trichomes, entire or broken; Longitudinally

Powdered Crude Drug Microscopy of Leaves and Barks. https://doi.org/10.1016/B978-0-12-818092-1.00044-0

cut fragments of spiral as well as annular xylem vessels. The general as well as specific features of *S.indicus* leaf powder are illustrated in Table 6.20.1 and-Figs. 6.20.2, 6.20.3A-E, 6.20.4, 6.20.5, 6.20.6 and 6.20.7 respectively.

TABLE 6.20.1 Specific as well as general characters of leaf powder of *Sphaeranthus indicus.*

Features		
Epidermis	**Trichomes**	**Calcium oxalate crystals**
Wavy	Abundant	Prismatic
Polygonal	Simple	
Stomata	Covering	
Sunken	Multicellular	**Cell contents**
	Long	Brownish matter
Xylem vessels	Curved	
Spiral	Cystolithic	
Annular	Swollen bases	

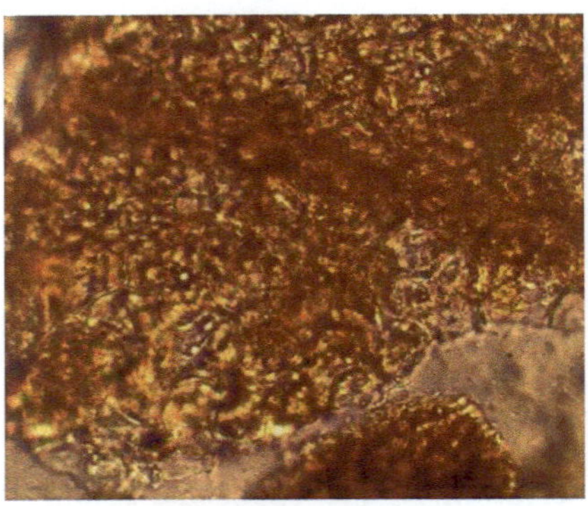

FIG. 6.20.2 Sunken stomata present in *Sphaeranthus indicus* leaf powder.

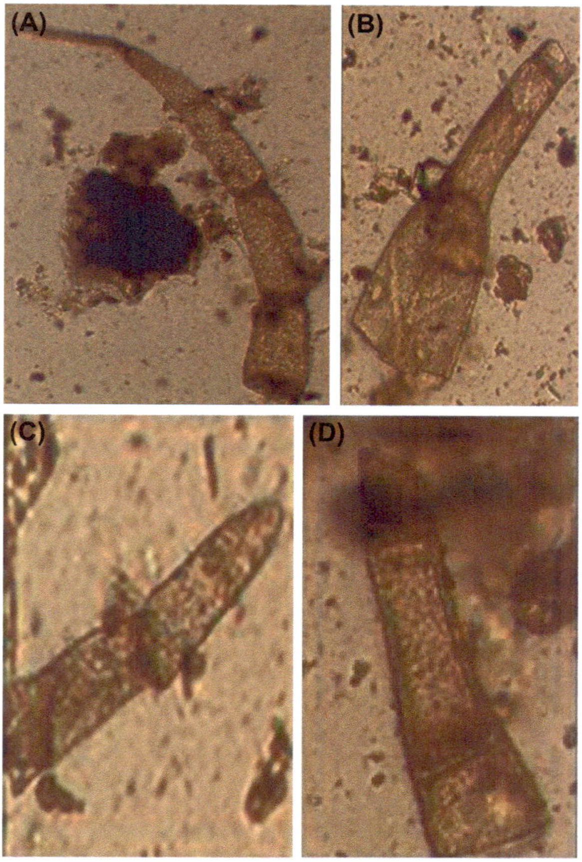

FIG. 6.20.3 (A–D) – Covering trichomes present in *Sphaeranthus indicus* leaf powder.

FIG. 6.20.4 Spiral xylem vessels present in *Sphaeranthus indicus* leaf powder.

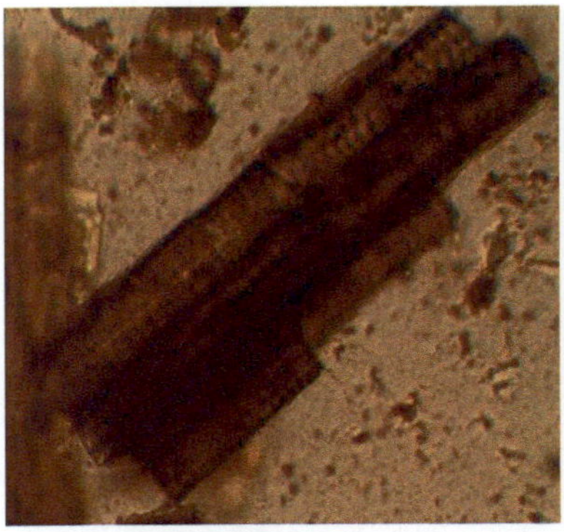

FIG. 6.20.5 Annular xylem vessels present in *Sphaeranthus indicus* leaf powder.

FIG. 6.20.6 Prismatic crystals present in *Sphaeranthus indicus* leaf powder.

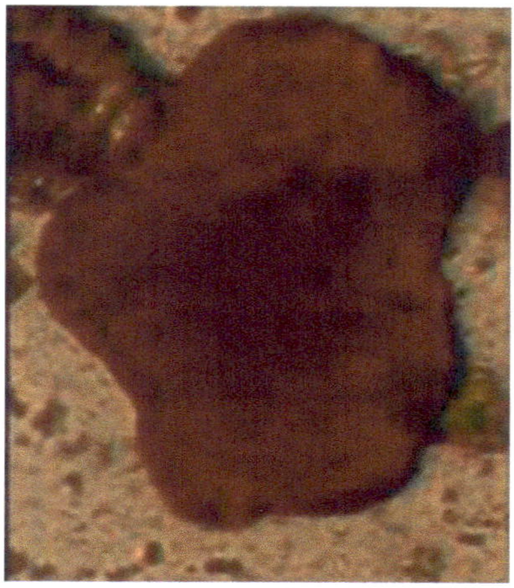

FIG. 6.20.7 Brownish matter present in *Sphaeranthus indicus* leaf powder.

Reference

[1] Government of India, Department of AYUSH, Ayurvedic Pharmacopoeia of India. Part I, vol. III, Ministry of Health and welfare, Ghaziabad, 1999-2017, p. 128.

Chapter 7

Microscopy/Monograph of adulterants and substitutes of powdered bark and leaf drugs

Adulteration and substitution of herbal drugs can cause a serious health risk to consumers. Based on the powder microscopic evaluation, specific characters can be identified which can differentiate between a genuine drug and substitute/adulterant. The following bark drugs along with their substitute/adulterant are selected for powder microscopic study:

7.1 *Albizzia lebbeck* (L.) Benth., Albizzia marginata (Lam.) Merr., Albizzia odoratissima (L.f.) Benth.
7.2 *Bauhinia variegata* L., Bauhinia malabarica Roxb., Bauhinia racemosa Vahl
7.3 *Dalbergia sissoo* DC., Dalbergia latifolia Roxb.
7.4 Holarrhena *antidysenterica* (Roth) Wall. Ex A. DC., *Wrightia tinctoria* R. Br.
7.5 *Saraca asoca* (Roxb.) Willd., *Polyalthia longifolia* (Sonn.) Thwaites
7.6 *Terminalia arjuna* (Roxb. ex DC.) Wight & Arn., *Terminalia bellirica* (Gaertn.) Roxb., *Terminalia calamansi* Rolfe, *Terminalia chebula* Retz.

The following leaf drugs along with their substitute/adulterant are selected for powder microscopic study:

7.7 *Adhotoda* vasica Nees, *Adenanthera pavonina* L., *Ailanthus excelsa* Roxb.
7.8 *Azadirachta indica* A. Juss., *Melia azadirach* L.
7.9 Cassia angustifolia M. Vahl, Cassia auriculata L., *Ailanthus* glandulosa Desf.
7.10 *Indigofera tinctoria* L., *Indigofera* arrecta A.Rich.
7.11 *Phyllanthus* fraternus G.L.Webster, *Phyllanthus* amarus Schumach. & Thonn.
7.12 *Vitex negundo* L. *Vitex trifolia* L.

Bark drugs

Chapter 7.1

Albizzia lebbeck – *Albizzia odoratissima* and *Albizzia marginata*

Albizzia lebbeck is a deciduous, perennial medium-sized legume tree commonly used in traditional medicine. There are more than 150 species in the genus Albizzia. *A.odoratissima* and *A.marginata* are commonly substituted with *A.lebbeck.*

A. lebbeck (L.) Benth.	Mimosaceae
Albizzia odoratissima (L.f.) Benth.	Fabaceae
Albizzia marginata (Lam.) Merr.	Mimosaceae

Common vernacular names: Bengali – Sirish, Siris; English – Siris Tree, Lebbeck Tree; Gujrati – Shirish; Hindi: Siris, Shiris; Kannada – Bagey, Bage Mara, Hombage; Malayalam – Vaka, Nanmenivaka; Marathi – Siris; Oriya – Sersuan, Sirisha; Punjabi – Sirish, Sareehn; Tamil – Vakai; Telugu – Dirisena [1].

Major chemical constituents: D-catechin, isomers of leucocyanidin and melacacidin, leucoanthrocyanidin, lebbecacidin, friedelin, β-sitosterol [2].

Traditional uses: Bronchitis, leprosy [1].

Macroscopic description: The morphological characters of *A. lebbeck – A.odoratissima and A.marginata* are illustrated in Table 7.1.1 and Figs. 7.1.1–7.1.3 respectively.

7.1.1 Powder microscopic description

A. lebbeck bark powder is greyish brown in colour showing abundant fragments of the cork cells - polygonal to hexagonal, often slightly elongated, thick or thin walled, with big oval to spherical, rounded cavities formed by disintegration of the parenchymatous cells; Prismatic crystals of calcium oxalate; Fragments of tangentially and radially longitudinally cut medullary rays and devoid of stone cells and sclereids.

A. odoratissima bark powder is blackish grey in colour showing fragments of the cork cells - polygonal to pentagonal in shape; Rosette crystals of calcium oxalate and isolated or group of stone cells oval in shape.

Powdered Crude Drug Microscopy of Leaves and Barks. https://doi.org/10.1016/B978-0-12-818092-1.00045-2

TABLE 7.1.1 Morphological characters of three species of *Albizzia*.

S. NO.	Features	*Albizzia lebbeck*	*Albizzia odoratissima*	*Albizzia marginata*
1.	Shape	Thick and flat pieces	Flat pieces	Flat pieces
2	Outer bark	Dark greyish brown, rough due to longitudinal and transverse cracks	Dark blackish grey	Smooth and grey in colour
3	Inner bark	Light yellow to grey and fibrous	Light yellowish brown and longitudinally striated	Light brown and longitudinally striated
4	Fracture	Laminated in outer region and fibrous in inner region	Short and Fibrous	Fibrous
5	Odour	Characteristic	Characteristic	Characteristic

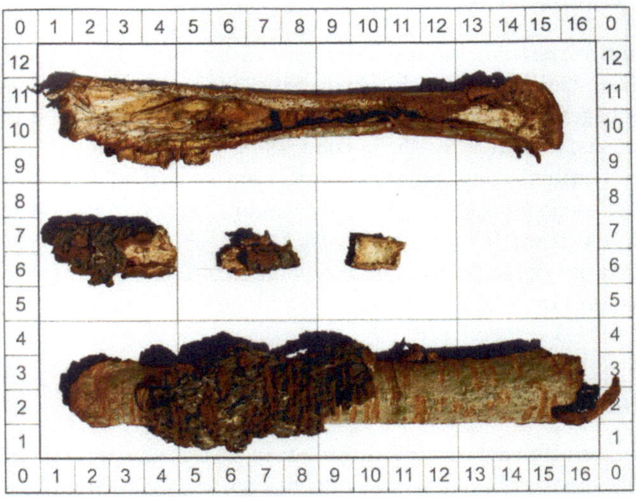

FIG. 7.1.1 *Albizzia lebbeck* bark.

A. marginata bark powder is light greyish in colour showing fragments of the cork cells pentagonal in shape; Prismatic crystals of calcium oxalate; Group of abundant stone cells smaller in size as compared to *A. odoratissima* and fragments of tangentially and radially longitudinally cut medullary rays.

FIG. 7.1.2 *Albizzia odoratissima* bark.

FIG. 7.1.3 *Albizzia marginata* bark.

Based upon the powder microscopic features, the bark powders of *A. lebbeck, A. odoratissima and A. marginata* can be distinguished as shown in Table 7.1.2 and Figs. 7.1.4–7.1.11 respectively. *A.lebbeck* and *A.marginata* bark powder shows prismatic crystals of calcium oxalate whereas *A.odoratissima* contained rosette crystals. Cork cells of *A.lebbeck* shows cval or

TABLE 7.1.2 Specific comparative feature of powdered bark powders of *Albizzia lebbeck, Albizzia odoratissima and Albizzia marginata.*

S.No.	*Features*	*Albizzia lebbeck*	*A.odoratissima*	*A.marginata*
1	**Common name**	Shiris	Kali Shiris	Silk tree
2	**Calcium oxalate crystals**	Prismatic	Rosette	Prismatic
3	**Stone cells**			
	Occurrence	NA	Isolated	Group
			Group	
	Walls	NA	Thin walled	Thick walled
	Shape	NA	Oval	Circular
				Oval
	Lumen	NA	Wide	Narrow
4	**Cork**	Oval	Thick walled	Thick walled
		Rounded	Polygonal	Pentagonal
			Pentagonal	

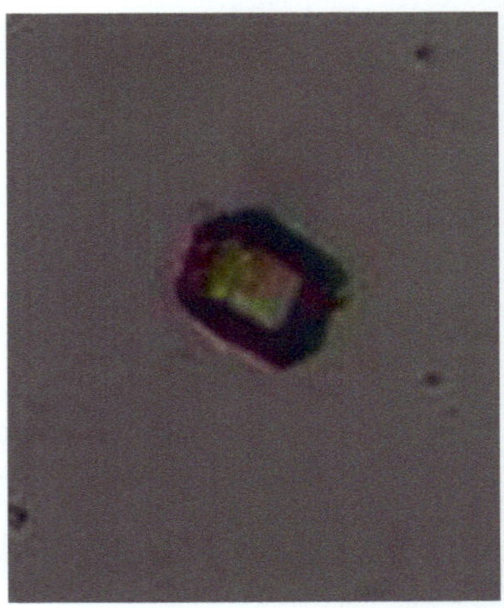

FIG. 7.1.4 Prismatic crystal from *Albizzia lebbeck* bark powder.

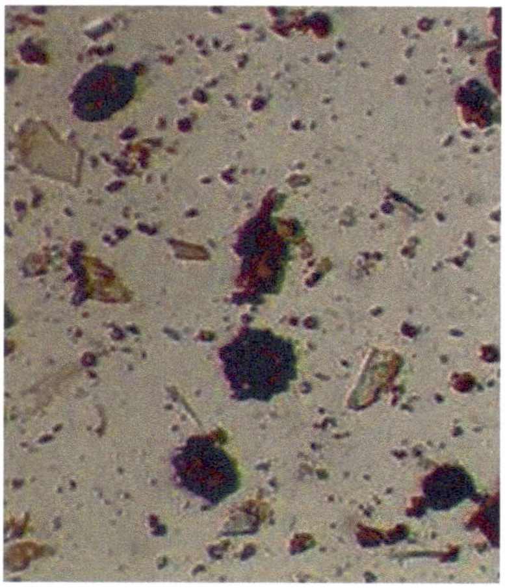

FIG. 7.1.5 Rosette crystal from *Albizzia odoratissima* bark powder.

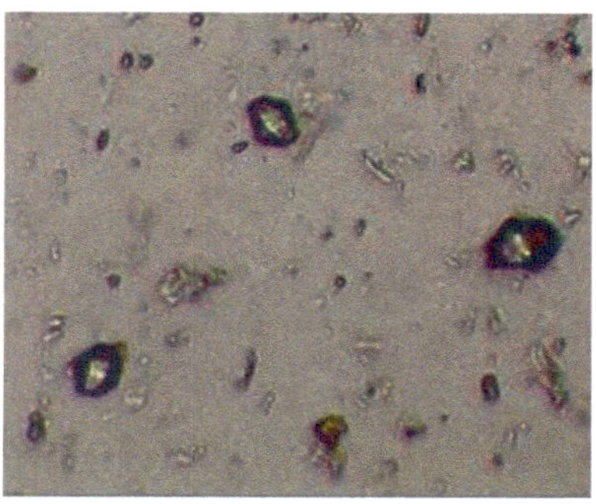

FIG. 7.1.6 Prismatic crystal from *Albizzia marginata* bark powder.

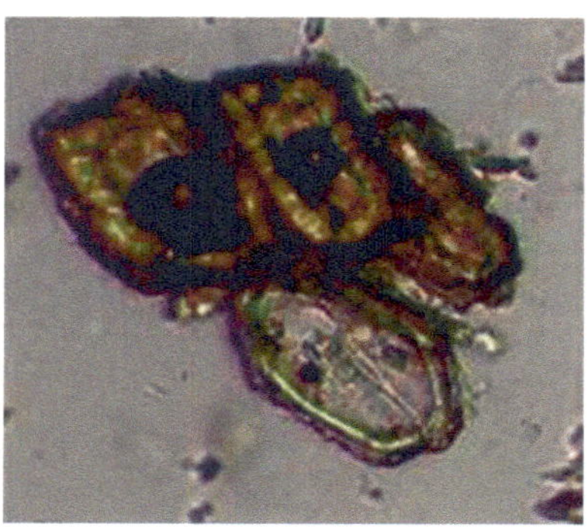

FIG. 7.1.7 Stone cells from *Albizzia odoratissima* bark powder.

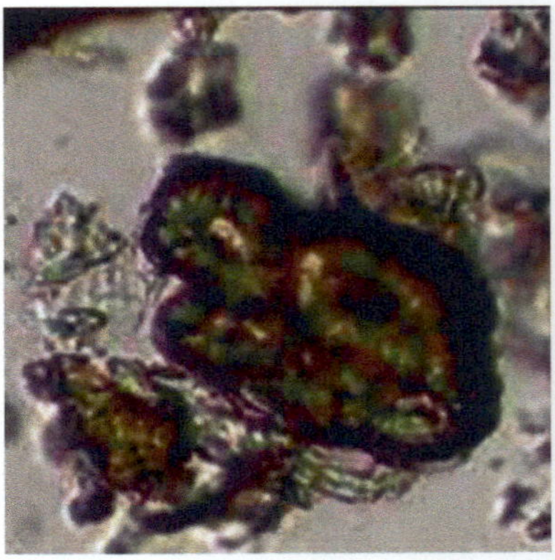

FIG. 7.1.8 Stone cells from *Albizzia marginata* bark powder.

FIG. 7.1.9 Cork cells from *Albizzia lebbeck* bark powder.

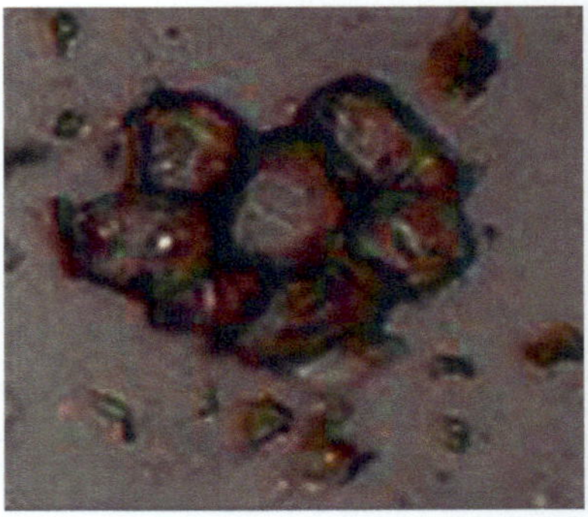

FIG. 7.1.10 Cork cells from *Albizzia odoratissima* bark powder.

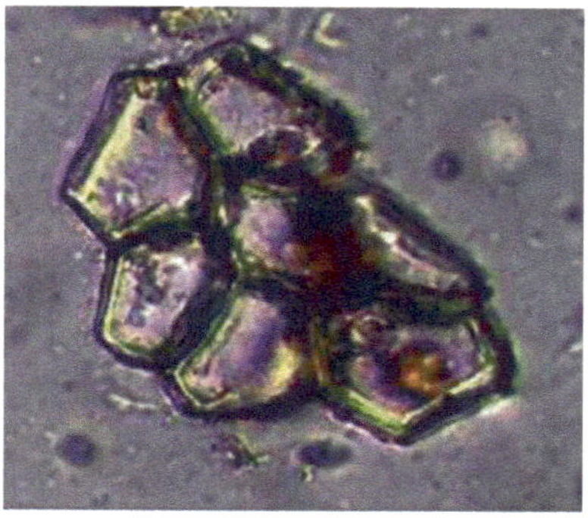

FIG. 7.1.11 Cork cells from *Albizzia marginata* bark powder.

rounded cavitation which is the most distinguishing character for this bark powder. The stone cells are thick walled with narrow lumen in A.*marginata* and thin walled with wide lumen in *A.odoratissima*. Stone cells are found to be absent in *A.lebbeck*.

References

[1] Government of India, Department of AYUSH, Ayurvedic Pharmacopoeia of India. Part I, vol. III, Ministry of Health and welfare, Ghaziabad, 1999-2017, p. 201.

[2] A.K. Gupta, N. Tandon, M. Sharma, Quality Standards of Indian Medicinal Plants, vol. 2, Medicinal plant unit, ICMR, New Delhi, India, 2008, p. 01.

Chapter 7.2

Bauhinia variegata – *Bauhinia racemosa* and *Bauhinia malabarica*

Bauhinia variegata is a species of flowering plant in the legume family commonly grown as ornamental tree. There are more than 500 species in the genus *Bauhinia. B.racemosa* and *B.malabarica* are other two important species of *Bauhinia* with medicinal value.

B. variegata L.	Fabaceae
Bauhinia racemosa Vahl	Fabaceae
Bauhinia malabarica Roxb.	Caeselpiniaceae

Common vernacular names: Assamese – Kancan, Kanchan; Bengali – Kanchana, Rakta Kanchana; English – Mountain Ebony; Gujarati – Champakati, Kanchnar, Kachnar; Hindi – Kachanar, Kanchanar, Kachnar; Kannada – Keyumandar, Kanchavala; Kashmiri – Kalad; Malayalam – Chuvanna Mandharam; Marathi – Kanchana, Raktakancana; Oriya – Kachana, Kaniara; Punjabi – Kanchnar; Tamil – Sigappu mandarai, Sihappu mantarai; Telugu – Deva Kanchanam [1].

Major chemical constituents: Lupeol, Naringenin, friedelin, β-sitosterol [2].

Traditional uses: Anthelmentic [2].

Macroscopic description: The morphological characters of *B. variegata* – *B. racemosa* and *B. malabarica* are illustrated in Table 7.2.1 and Figs. 7.2.1–7.2.3 respectively.

7.2.1 Powder microscopic description

B. variegata bark powder is reddish brown in colour showing plenty of stone cells, most of them are circular to oval in shape, with wide, pitted lumen and striated and beaded walls; Sclereids few, oval to rectangular, with wide, pitted lumen and striated and pitted walls; Fragments of tangentially and radially longitudinally cut medullary rays; Abundant prismatic crystals and fragments of polygonal to hexagonal cork cells [3].

Powdered Crude Drug Microscopy of Leaves and Barks. https://doi.org/10.1016/B978-0-12-818092-1.00046-4

TABLE 7.2.1 Morphological characters of three species of *Bauhinia*.

S. No.	Features	*Bauhinia variegata*	*Bauhinia racemosa*	*Bauhinia malabarica*
1.	Shape	Curved or channeled	Curved or channeled	Curved or channeled
2	Outer bark	Dark brown with transverse and longitudinal cracks	Greyish brown externally	Dark greyish brown, rough and peeling in linear flakes
3	Inner bark	Reddish, hard, smooth to faintly longitudinally striated	Light brown	Light reddish brown and smooth
4	Fracture	Outer short, inner fibrous	Outer short, inner fibrous	Fibrous
5	Odour	Not characteristic	Characteristic	Characteristic

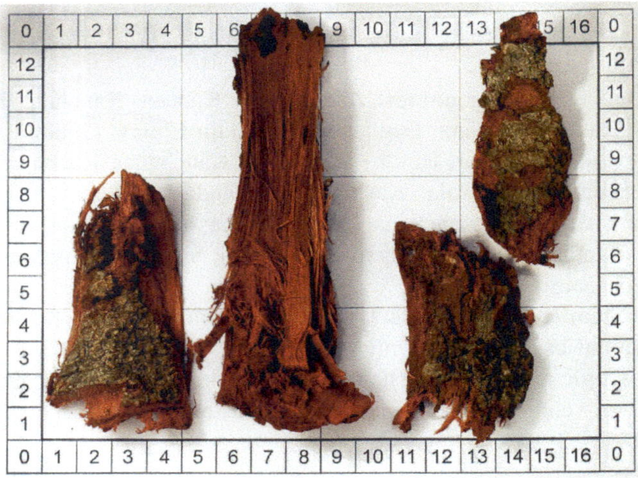

FIG. 7.2.1 *Bauhinia variegata* bark.

B. racemosa bark powder is light brown in colour showing few round to oval small stone cells; Polygonal to hexagonal cork cells with yellowish brown content; Fragments of tangentially and radially longitudinally cut medullary rays and prismatic crystals, less abundant as compared to *B. variegata*.

B. malabarica bark powder is light reddish brown showing prismatic, rosette and cluster crystals of calcium oxalate; Pentagonal to hexagonal cork

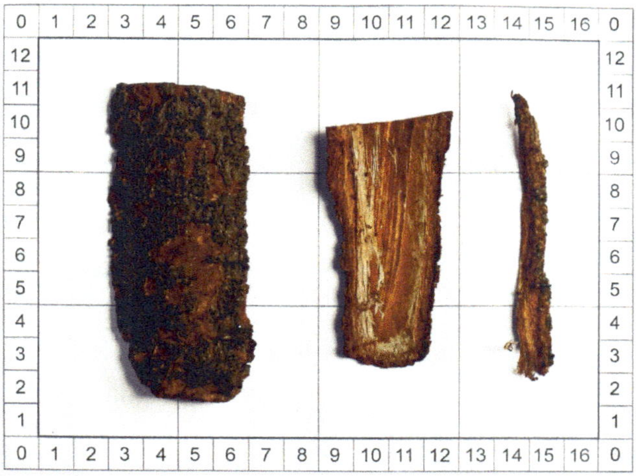

FIG. 7.2.2 *Bauhinia racemosa* bark.

FIG. 7.2.3 *Bauhinia malabarica* bark.

cells and simple septate fibres and very few, occasional stone cells oval in shape.

Based upon the powder microscopic features, the bark powders of *B. variegata, B. racemosa* and *B. malabarica* can be distinguished as shown in Table 7.2.2 and Figs. 7.2.4–7.2.15. Sclereids were found to be absent in *B. racemosa and B. malabarica* bark powder. On the other hand, few oval and rectangular sclereids were present in *B. variegata*. The size variation of stone cells in *B. variegata, B. racemosa and B. malabarica* was found to be

TABLE 7.2.2 Specific comparative feature of powdered bark powders of *Bauhinia variegata, Bauhinia racemosa* and *Bauhinia malabarica.*

S. No.	Features	*Bauhinia variegata*	*Bauhinia racemosa*	*Bauhinia malabarica*
1	**Common name**	Kachanar	Katmauli	Amli
2	**Stone cells**			
	Occurrence	Abundant	Few	Few
		Isolated	Isolated	Isolated
		Group	Group	
	Shape	Circular	Oval	Oval
		Oval	Triangular	
	Walls	Striated	Striated	Striated
		Beaded		
	Lumen	Wide	Narrow	Wide
		Pitted		
	Size	40–55 μm	65 μm	20–40 μm
3	**Sclereids**			
	Occurrence	Few	NA	NA
	Shape	Oval	NA	NA
		Rectangular		
	Walls	Striated	NA	NA
		Pitted		
	Lumen	Wide	NA	NA
		Pitted		
4	**Calcium oxalate crystals**	Prismatic	Prismatic	Prismatic
				Rosette
				Cluster
5	**Cork**	Polygonal	Polygonal	Pentagonal
		Hexagonal	Hexagonal	Hexagonal
			Yellowish Brown content	

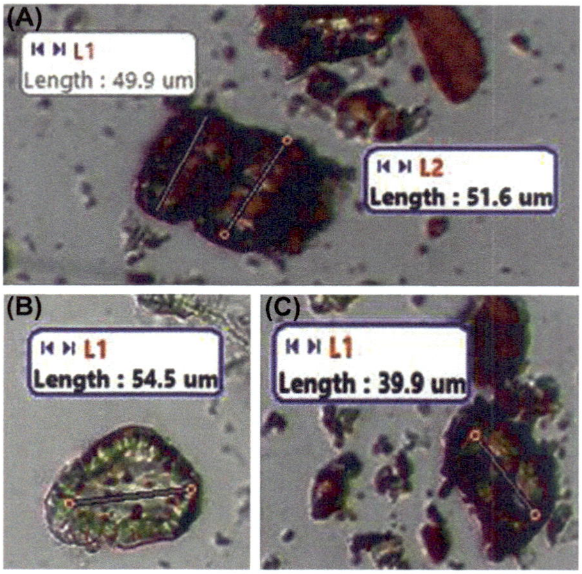

FIG. 7.2.4 (A–C) – Stone cells from *Bauhinia variegata* bark powder.

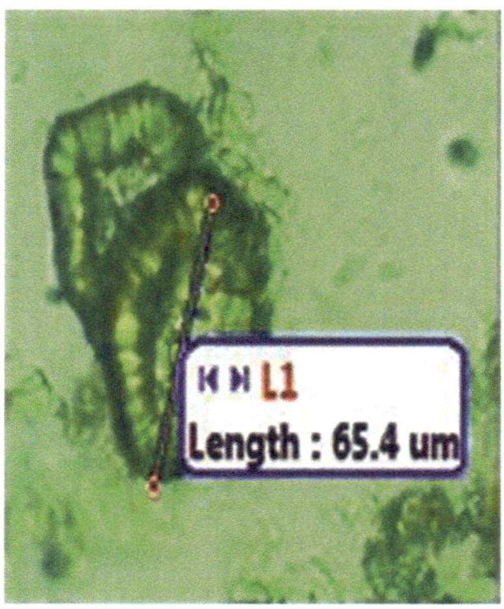

FIG. 7.2.5 Stone cells from *Bauhinia racemosa* bark powder.

FIG. 7.2.6 Stone cells from *Bauhinia malabarica* bark powder.

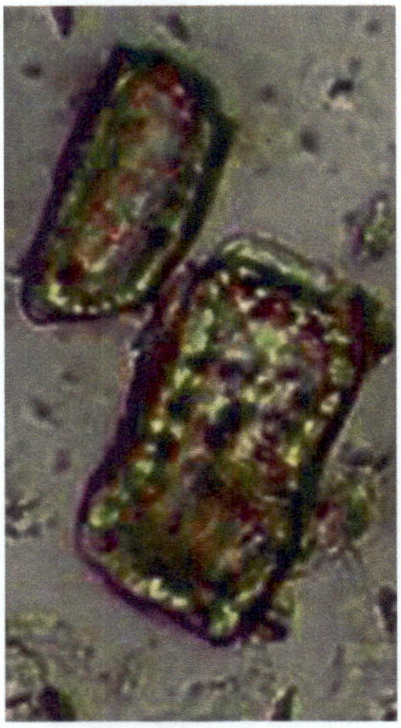

FIG. 7.2.7 Sclereids from *Bauhinia variegata* bark powder.

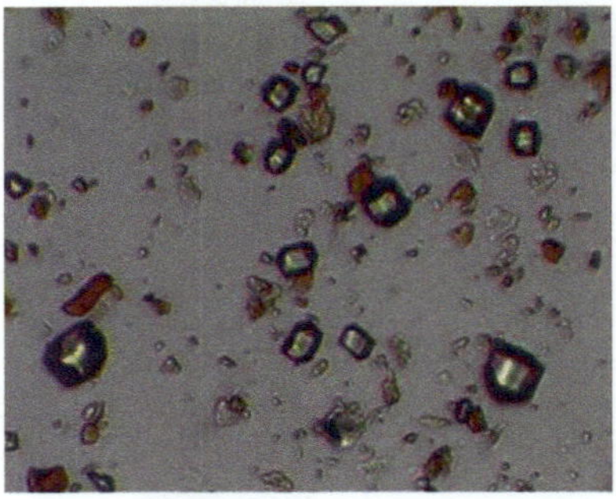

FIG. 7.2.8 Prismatic crystals from *Bauhinia variegata* bark powder.

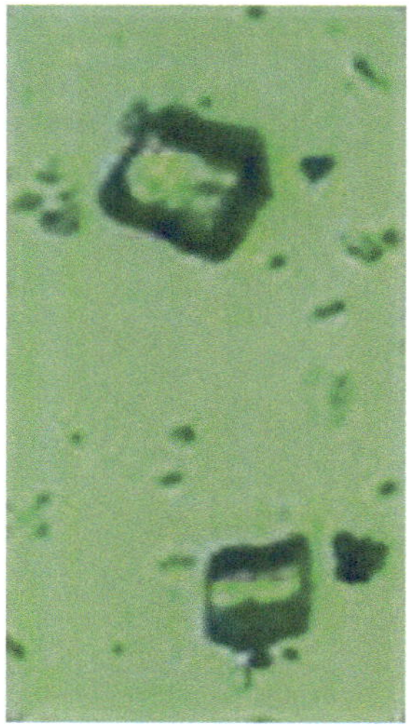

FIG. 7.2.9 Prismatic crystals from *Bauhinia racemosa* bark powder.

FIG. 7.2.10 Prismatic crystals from *Bauhinia malabarica* bark powder.

40–55 μm, 65 μm and 20–40 μm respectively. The cork cells of *B. racemosa* were rich in yellowish brown content as compared to other two species. Powder microscopy revealed the presence of prismatic crystals in all the three species of *Bauhinia*, but rosette and cluster crystals were present in *B. racemosa*.

FIG. 7.2.11 Rosette crystals from *Bauhinia malabarica* bark powder.

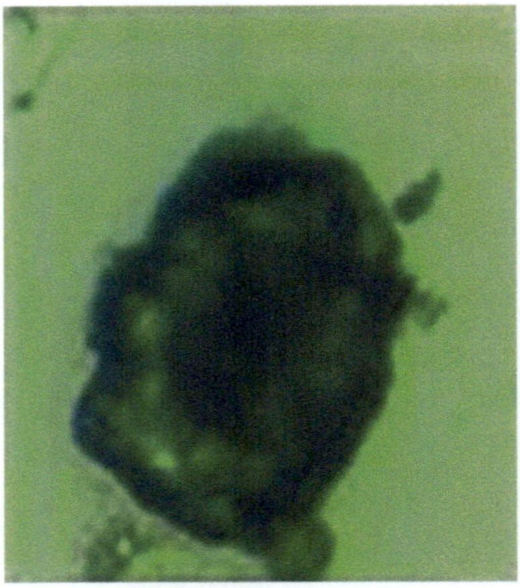

FIG. 7.2.12 Cluster crystals from *Bauhinia malabarica* bark powder.

FIG. 7.2.13 Cork cells from *Bauhinia variegata* bark powder.

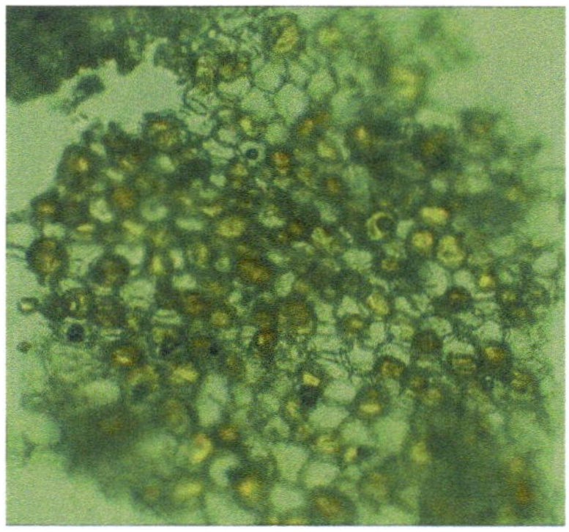

FIG. 7.2.14 Cork cells from *Bauhinia racemosa* bark powder.

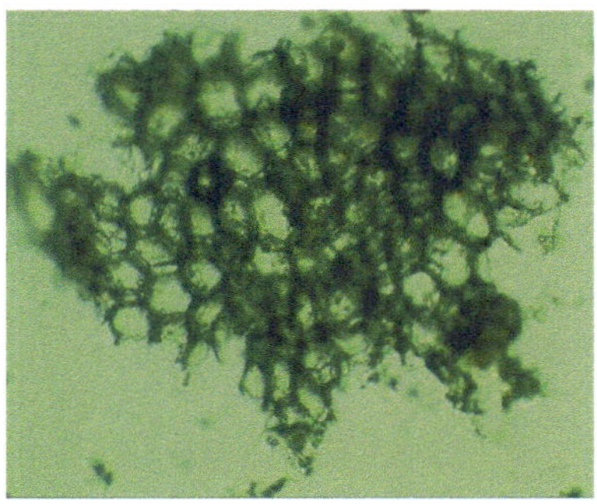

FIG. 7.2.15 Cork cells from *Bauhinia malabarica* bark powder.

References

[1] Government of India, Department of AYUSH, Ayurvedic Pharmacopoeia of India. Part I, vol. I, Ministry of Health and Welfare, Ghaziabad, 1999–2017, p. 74.

[2] A.K. Gupta, N. Tandon, M. Sharma, Quality Standards of Indian Medicinal Plants, vol. 5, Medicinal Plant Unit, ICMR, New Delhi, India, 2008, p. 118.

[3] M.G. Chauhan, A.P.G. Pillai, Microscopic Profile of Powdered Drugs Used in Indian System of Medicine, Bark Drugs, vol. 1, Institute of Ayurvedic Medicinal Plant Sciences, Jamnagar, Gujarat, India, 2005, p. 50.

Chapter 7.3

Dalbergia sissoo and *Dalbergia latifolia*

Dalbergia is a large genus of small to medium-size trees, shrubs and lianas in the pea family, Fabaceae. *Dalbergia sissoo* is an Indian variety which yields hard inexpensive timber as compared to *Dalbergia latifolia.*

D. sissoo DC. Fabaceae
D. latifolia Roxb. Fabaceae

Common vernacular names: Bengali – Shishu; English – Sissoo Tree; Gujarati – Siram; Hindi – Seesam; Kannada – Eragundimavu, Bindi; Malayalam – Irupoola; Marathi – Sisu, Shisav; Oriya – Sisu, Sinsapa; Punjabi – Sheesham; Tamil – Irupoolai; Telugu – Irugudu, Virugudu, Sissoo [1].

Major chemical constituents: Dalberginone, Dalbergin, Methyldalbergin, 4-Phenylchromene, Dalbergichromene and Isotectorigenin [1].

Traditional uses: Leprosy and skin diseases [1].

Macroscopic description: The morphological characters of *D. sissoo* and *D. latifolia* are illustrated in Table 7.3.1 and Figs. 7.3.1 and 7.3.2 respectively.

TABLE 7.3.1 Morphological characters of *Dalbergia sissoo* and *Dalbergia latifolia.*

S. No.	Features	*Dalbergia sissoo*	*Dalbergia latifolia*
1	Shape	Long, curved or flat	Flat pieces
2	Outer bark	Dark reddish brown, rough with broad and shallow longitudinal fissures	Light reddish brown in colour and longitudinal fissures
3	Inner bark	Yellowish brown, longitudinally striated	Reddish brown, longitudinally striated
4	Fracture	Fibrous	Fibrous
5	Odour	Characteristic	Slight

Powdered Crude Drug Microscopy of Leaves and Barks. https://doi.org/10.1016/B978-0-12-818092-1.00047-6

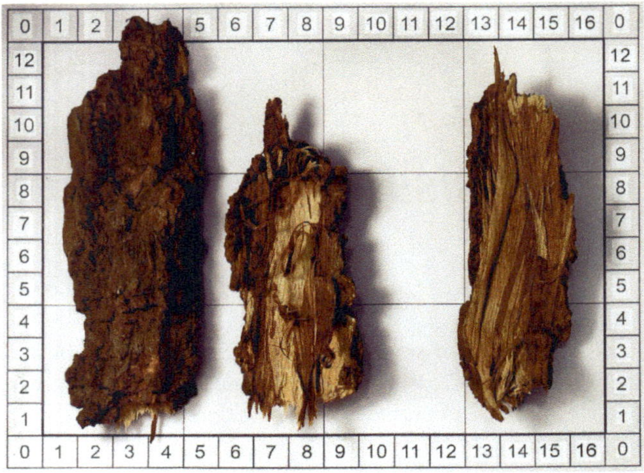

FIG. 7.3.1 *Dalbergia sissoo* bark.

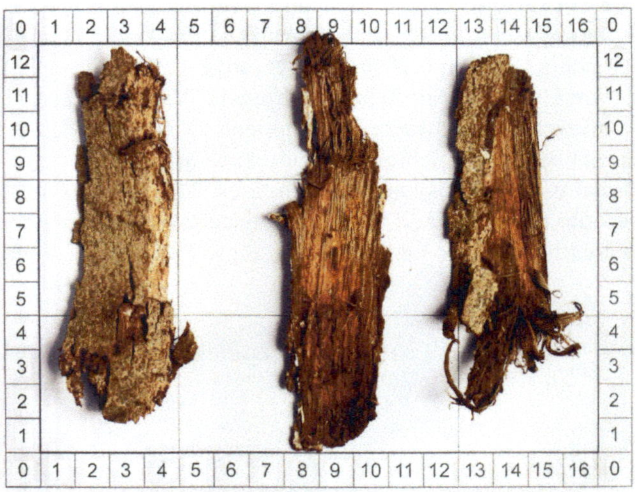

FIG. 7.3.2 *Dalbergia latifolia* bark.

7.3.1 Powder microscopic description

D. sissoo bark powder is dark reddish brown in colour showing abundant prismatic crystals of calcium oxalate scattered throughout the powder; Fragments of tangentially and radially longitudinally cut medullary rays; Hexagonal to pentagonal, thick walled cork cells and abundant starch grains, spherical to oval in shape.

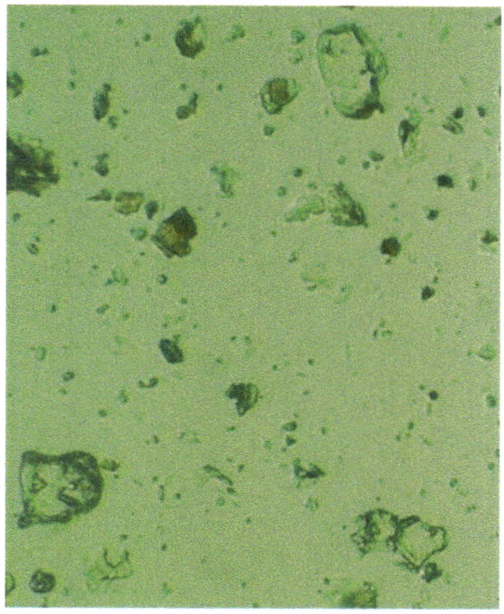

FIG. 7.3.3 Stone cells from *Dalbergia latifolia* bark powder.

D. latifolia bark powder is light reddish brown in colour showing few prismatic and rosette crystals of calcium oxalate scattered throughout the field; Fragments of tangentially and radially longitudinally cut medullary rays; Hexagonal, polygonal to pentagonal, thick walled cork cells; Abundant starch grains, spherical to oval in shape; Fragments of simple fibres and Oval to pentagonal stone cells with wide lumen.

Based upon the powder microscopic features, the bark powders of *D. sissoo* and *D. latifolia* can be distinguished as shown in Table 7.3.2 and

TABLE 7.3.2 Specific comparative feature of powdered bark powders of *Dalbergia sissoo* and *Dalbergia latifolia*.

S. No.	Features	*Dalbergia sissoo*	*Dalbergia latifolia*
1	**Common name**	Seesam	Indian Rosewood
2	**Stone cells**		
	Occurrence	NA	Few
			Isolated
			Group
	Shape	NA	Oval
			Pentagonal
	Walls	NA	Thin walled
	Lumen	NA	Wide

Continued

TABLE 7.3.2 Specific comparative feature of powdered bark powders of *Dalbergia sissoo* and *Dalbergia latifolia*.—cont'd

S. No.	Features	*Dalbergia sissoo*	*Dalbergia latifolia*
3	**Calcium oxalate crystals**	Prismatic	Prismatic
			Rosette
4	**Cork**	Hexagonal	Hexagonal
		Pentagonal	Polygonal
			Pentagonal
5	**Fibres**		
	Occurrence	NA	Isolated
			Few
	Walls	NA	Thin walled
			Septate
	Shape	NA	Very long
			Pointed apex
	Lumen	NA	Narrow

Figs. 7.3.4–7.3.9. *D. sissoo* bark powder was deficient in stone cells and fibres but *D. latifolia* showed few, oval and pentagonal stone cells along with very long, septate fibres with pointed apex. *D. latifolia* bark powder consisted of rosette crystals as compared to prismatic crystals found in *D. sissoo*.

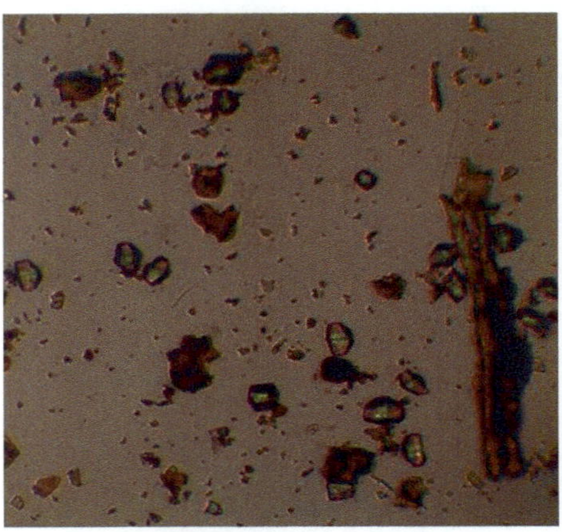

FIG. 7.3.4 Prismatic crystals from *Dalbergia sissoo* bark powder.

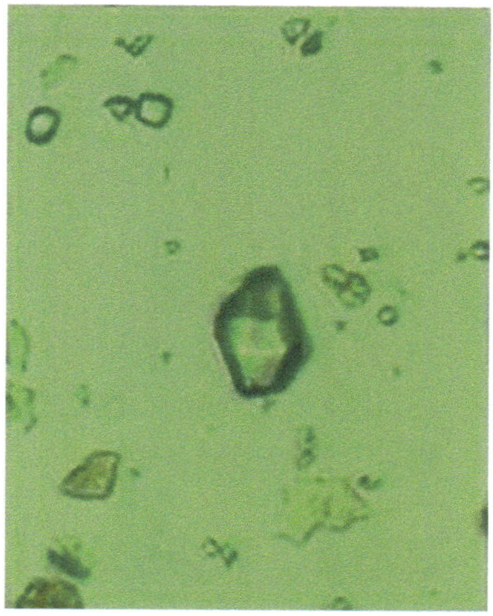

FIG. 7.3.5 Prismatic crystals from *Dalbergia latifolia* bark powder.

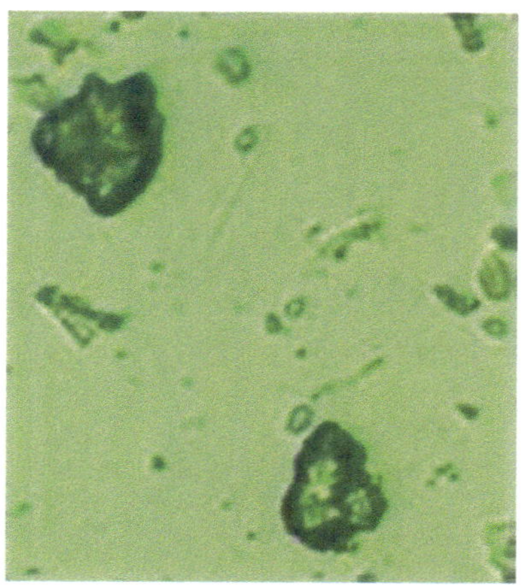

FIG. 7.3.6 Rosette crystals from *Dalbergia latifolia* bark powder.

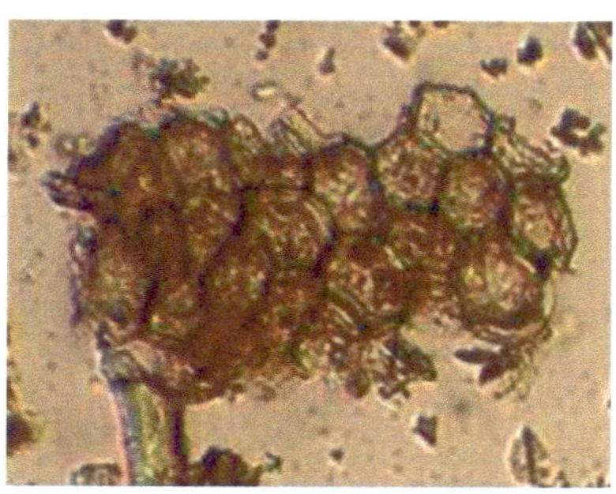

FIG. 7.3.7 Cork cells from *Dalbergia sissoo* bark powder.

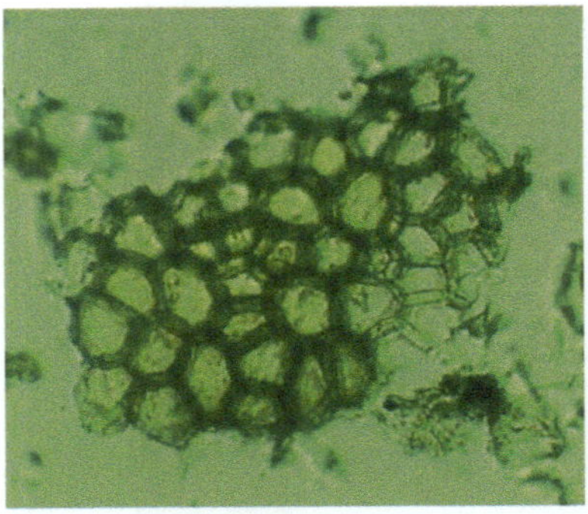

FIG. 7.3.8 Cork cells from *Dalbergia latifolia* bark powder.

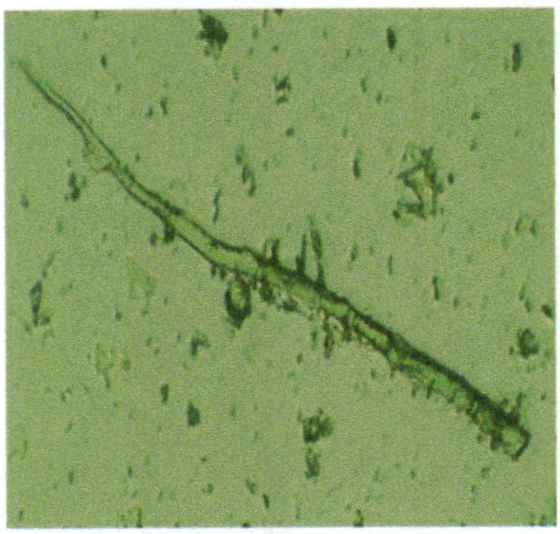

FIG. 7.3.9 Simple fibre from *Dalbergia latifolia* bark powder.

Reference

[1] Government of India, Department of AYUSH, Ayurvedic Pharmacopoeia of India. Part I, vol. III, Ministry of Health and Welfare, Ghaziabad, 1999–2017, p. 200.

Chapter 7.4

Holarrhena antidysentrica and *Wrightia tinctoria*

Kutaja is one among the most commonly used medicinal plants and having two varieties such as Pum (male) and Stree (female). The two source plants considered for Pum Kutaja and Stree Kutaja are *Holarrhena antidysenterica* and *Wrightia tinctoria* respectively, both belonging to family Apocynaceae. It is also known that the bark of *W. tinctoria* is commonly used as an adulterant of Kurchi Bark (*H. antidysenterica*).

Holarrhena antidysentrica (Roth) Wall. Ex A. DC	Apocynaceae
W. tinctoria R. Br	Apocynaceae

Common vernacular names: Assamese – Dudhkuri; Bengali – Kurchi; English – Ester tree, Conessi bark; Gujarati – Kuda, Kadachhal, Kudo; Hindi – Kurchi, Kuraiya; Kannada – Kodasige, Halagattigida, Halagatti Mara; Kashmiri – Kogad; Malayalam – Kutakappala; Marathi – Pandhra Kuda; Oriya – Kurei, Keruan; Punjabi – Kurasukk, Kura; Tamil – Kudasapalai; Telugu – Kodisapala, Palakodisa [1].

Major chemical constituents: Conessine and related alkaloids [1].

Traditional uses: Dysentery, amoebicidal, anthelmentic [1].

Macroscopic description: The morphological characters of *H. antidysentrica* and *W. tinctoria* are illustrated in Table 7.4.1 and Figs. 7.4.1 and 7.4.2 respectively.

7.4.1 Powder microscopic description

H. antidysentrica bark powder is buff brown in colour showing abundant stone cells, isolated or in fairly large groups of various size, shape, and thickness, rectangular to polygonal with strongly thickened pitted walls and some with striations; Occasionally elongated sclereids with very thick walls and narrow lumen; Prismatic crystals of calcium oxalate and oval to spherical simple starch grains scattered as such and fragments of thick walled, polygonal to somewhat rectangular cork cells [2].

Powdered Crude Drug Microscopy of Leaves and Barks. https://doi.org/10.1016/B978-0-12-818092-1.00048-8

TABLE 7.4.1 Morphological characters of *Holarrhena antidysentrica* and *Wrightia tinctoria*.

S. No.	Features	*Holarrhena antidysentrica*	*Wrightia tinctoria*
1	Shape	Channeled cut pieces and quills	Recurved or curved
2	Outer bark	Buff to brownish, longitudinally wrinkled	Dark brown and rough with irregular cracks
3	Inner bark	Brownish, rough and scaly	Cream in colour, rough due to closely arranged longitudinal ridges
4	Fracture	Short and granular	Outer short and inner fibrous
5	Odour	Not characteristic	Not characteristic

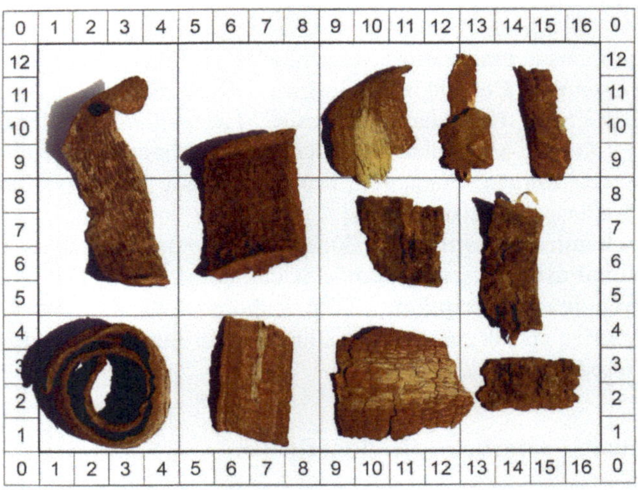

FIG. 7.4.1 *Holarrhena antidysentrica* bark.

W. tinctoria is dark brown in colour showing fragments of hexagonal to pentagonal, thick walled cork cells; Prismatic crystals of calcium oxalate and plenty of simple and compound starch grains; Fragments of radially and tangentially-longitudinally cut medullary rays and plenty of stone cells with wide or narrow lumen and distinct radiating walls.

FIG. 7.4.2 *Wrightia tinctoria* bark.

TABLE 7.4.2 Specific comparative feature of powdered bark powders of *Holarrhena antidysentrica* and *Wrightia tinctoria*.

S. No.	Features	*Holarrhena antidysentrica*	*Wrightia tinctoria*
1	**Common name**	Kurchi	Indrajan
2	**Stone cells**		
	Occurrence	Abundant	Few
		Isolated	Isolated
		Group	
	Shape	Oval	Oval
		Polygonal	Spherical
		Spherical	
		Rectangular	
	Walls	Thick walled	Thick walled
		Striations	Radiating pits
		Radiating pits	
	Lumen	Wide	Wide
		Narrow	Narrow
		Pitted	

Continued

TABLE 7.4.2 Specific comparative feature of powdered bark powders of *Holarrhena antidysentrica* and *Wrightia tinctoria*.—cont'd

S. No.	Features	*Holarrhena antidysentrica*	*Wrightia tinctoria*
3	**Sclereids**		
	Occurrence	Few	NA
		Isolated	
	Shape	Elongated	NA
		Pentagonal	
		Peg like extension	
	Wall	Thick walled	NA
	Lumen	Narrow	NA
		Pitted	
		Wide	

Based upon the powder microscopic features, the bark powders of *H. antidysentrica* and *W. tinctoria* can be distinguished as shown in Table 7.4.2 and Figs. 7.4.3–7.4.6. *H. antidysentrica* bark powder demonstrated wide variety of stone cells and sclereids whereas sclereids are absent and few, isolated stone cells are present in *W. tinctoria*.

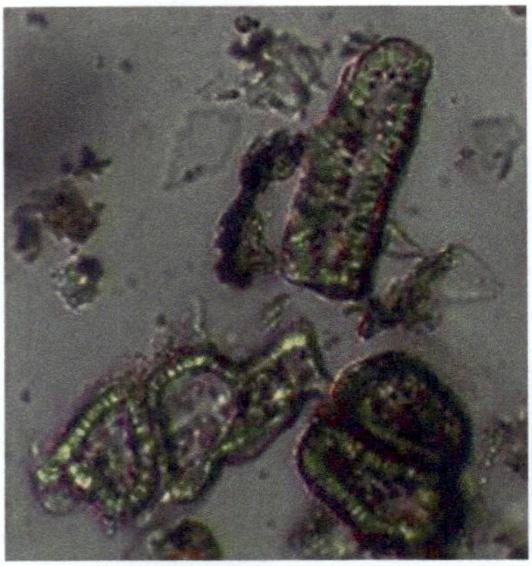

FIG. 7.4.3 Stone cells from *Holarrhena antidysentrica* bark powder.

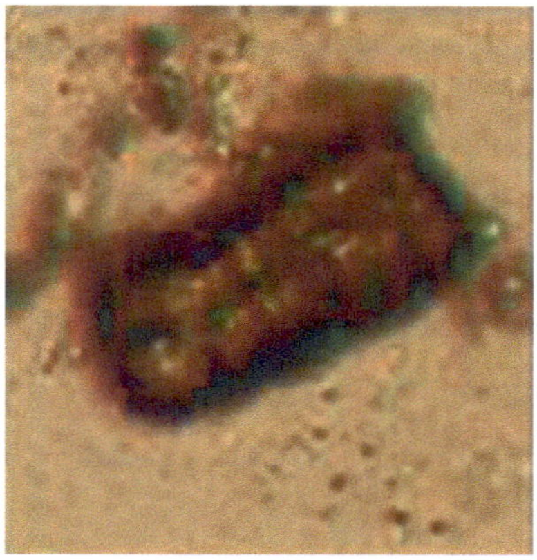

FIG. 7.4.4 Stone cells from *Wrightia tinctoria* bark powder.

FIG. 7.4.5 Stone cells from *Wrightia tinctoria* bark powder.

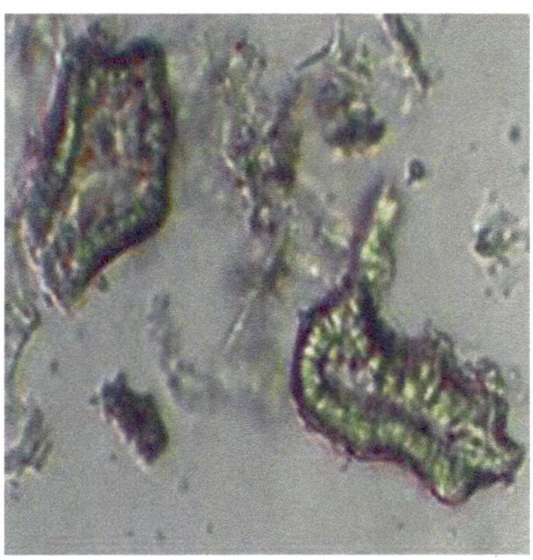

FIG. 7.4.6 Sclereids from *Holarrhena antidysentrica* bark powder.

References

[1] Government of India, Department of AYUSH, Ayurvedic Pharmacopoeia of India. Part I, vol. I, Ministry of Health and Welfare, Ghaziabad, 1999–2017, p. 78.

[2] M.G. Chauhan, A.P.G. Pillai, Microscopic Profile of Powdered Drugs Used in Indian System of Medicine, Bark Drugs, vol. 1, Institute of Ayurvedic Medicinal Plant Sciences, Jamnagar, Gujarat, India, 2005, p. 60.

Chapter 7.5

Saraca asoca and *Polyalthia longifolia*

Saraca asoca is an important indigenous medicinal plant used for the treatment of various female disorders. *Polyalthia longifolia* is the most common adulterant of *S. asoca*, also commonly known by the name Asokha.

S. asoca (Roxb.) Willd.	Caeselpiniaceae
P. longifolia (Sonn.) Thwaites	Annonaceae

Common vernacular names: Assamese – Ashoka; Bengali – Ashoka; English – Asok Tree; Gujarati – Ashoka; Hindi – Ashoka; Kannada – Ashokadamara, Ashokamara, Kankalimara; Kashmiri – Ashok; Malayalam – Asokam; Marathi – Ashok; Oriya – Ashoka; Punjabi – Asok; Tamil – Asogam, Asogu, Asokam; Telugu – Ashokapatta [1].

Major chemical constituents: Lignan glycosides [2].

Traditional uses: Uterine disorders [2].

Macroscopic description: The morphological characters of *S. asoca* and *P. longifolia* are illustrated in Table 7.5.1 and Figs. 7.5.1 and 7.5.2 respectively.

7.5.1 Powder microscopic description

S. asoca bark powder is brownish grey in colour. The tannin content of *S. asoca* bark powder is very high. Overnight soaking in water is not enough for visualization of characters. Therefore, it requires treatment with potassium chlorate. After treatment with oxidising agent bark powder shows abundant stone cells, squarish, oval to rectangular in shape with wide and narrow lumen and with distinct radiating pits and striated walls; Abundant fibrous sclereids, isolated or in groups, thick walled with distinct pits and striations; Polygonal to rectangular cork cells; Prismatic crystals of calcium oxalate and starch grains [3].

P. longifolia bark powder is brown in colour showing abundant, very long fibres with irregularly highly thickened walls occasionally exhibiting fine reticulations; Fragments of polygonal to pentagonal cork cells; Few stone cells, thick walled and polygonal in shape and abundant starch grains and micro-acicular crystals of calcium oxalate.

Powdered Crude Drug Microscopy of Leaves and Barks. https://doi.org/10.1016/B978-0-12-818092-1.00049-X

TABLE 7.5.1 Morphological characters of *Saraca asoca* and *Polyalthia longifolia*.

S. No.	Features	*Saraca asoca*	*Polyalthia longifolia*
1	Shape	Channeled	Thick and curved pieces
2	Outer bark	Brownish grey, cracked and transversely ridged	Brown, rough, ridged, ridges and furrows faint, lenticels vertical
3	Inner bark	Reddish brown with fine longitudinal strands and fibres	Yellowish brown, longitudinally striated
4	Fracture	Splintery	Hard and strongly fibrous
5	Odour	Characteristic	Pleasant

FIG. 7.5.1 *Saraca asoca* bark.

Based upon the powder microscopic features, the bark powders of *S. asoca* and *P. longifolia* can be distinguished as shown in Table 7.5.2 and Figs. 7.5.3–7.5.10. The bark powder of *S. asoca* showed wide range of stone cells and sclereids in terms of size, shape and abundance whereas polygonal stone cells are present in *P. longifolia* bark powder and completely devoid of sclereids. Compound starch grains with cleft hilum are seen in *P. longifolia* as compared to simple, spherical starch grains with centrally located hilum present in *S. asoca* bark powder.

FIG. 7.5.2 *Polyalthia longifolia* bark.

TABLE 7.5.2 Specific comparative feature of powdered bark powders of *Saraca asoca* and *Polyalthia longifolia*.

S. No.	Features	*Saraca asoca*	*Polyalthia longifolia*
1	**Common name**	Sita asokha	Debdari
2	**Stone cells**		
	Occurrence	Abundant	Few
		Isolated	Isolated
		Group	Group
	Shape	Rectangular	Polygonal
		Oval	
		Squarish	
	Walls	Radiating pits	Thick walled
		Striated walls	
	Lumen	Wide	Narrow
		Narrow	

Continued

TABLE 7.5.2 Specific comparative feature of powdered bark powders of *Saraca asoca* and *Polyalthia longifolia*.—cont'd

S. No.	Features	*Saraca asoca*	*Polyalthia longifolia*
2	**Sclereids**		
	Occurrence	Abundant	NA
		Isolated	
		Group	
	Shape	Oblong	NA
		Triangular	
	Walls	Fibrous	NA
		Thick walled	
		Distinct pits	
		Striations	
	Lumen	Wide	NA
3	**Cork**	Thick walled	Polygonal
		Polygonal	Pentagonal
		Yellowish brown	
		Rectangular	
4	**Starch grains**	Simple	Compound
		Spherical	Cleft hilum
		Oval	
		Centrally located hilum	

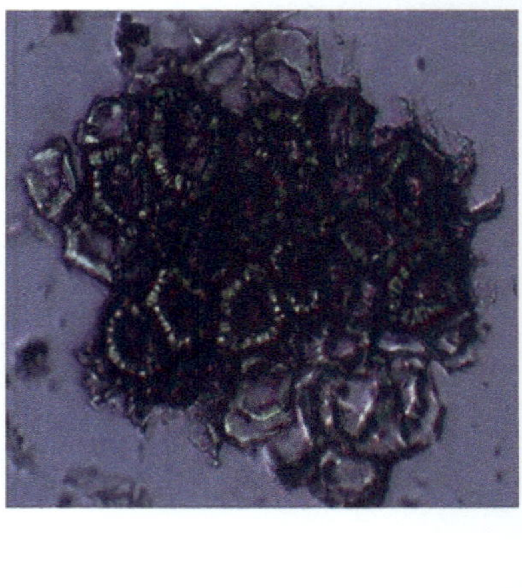

FIG. 7.5.3 Stone cells present in *Saraca asoca* bark powder.

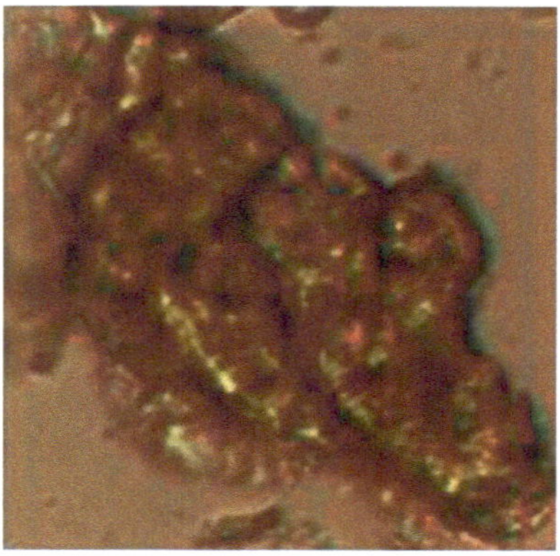

FIG. 7.5.4 Stone cells present in *Polyalthia longifolia* bark powder.

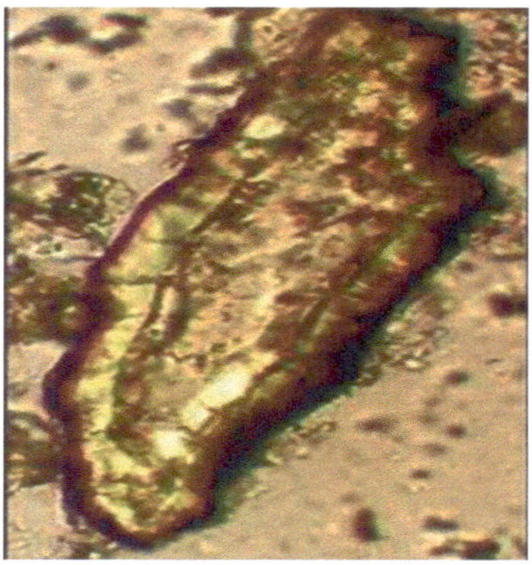

FIG. 7.5.5 Sclereid present in *Saraca asoca* bark powder.

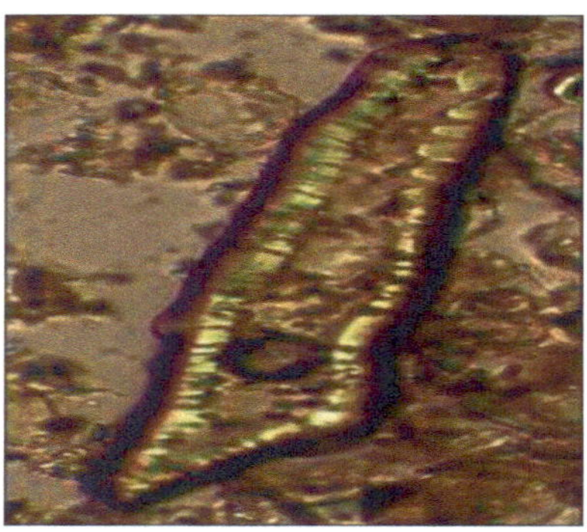

FIG. 7.5.6 Triangular sclereid present in *Saraca asoca* bark powder.

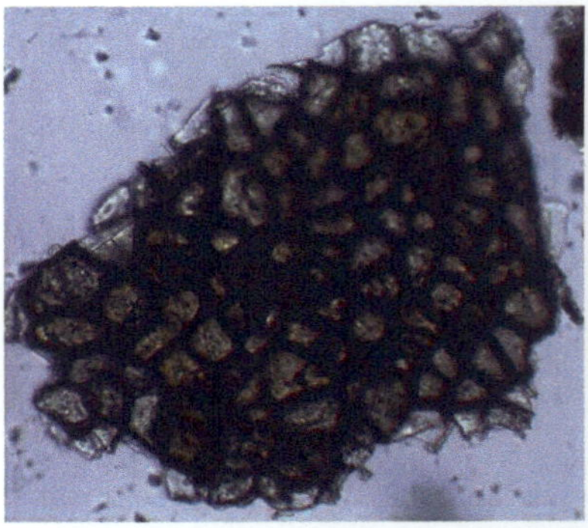

FIG. 7.5.7 Cork cells present in *Saraca asoca* bark powder.

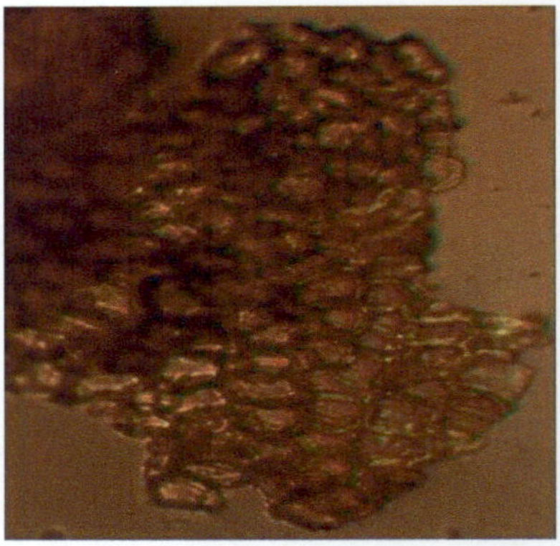

FIG. 7.5.8 Cork cells present in *Polyalthia longifolia* bark powder.

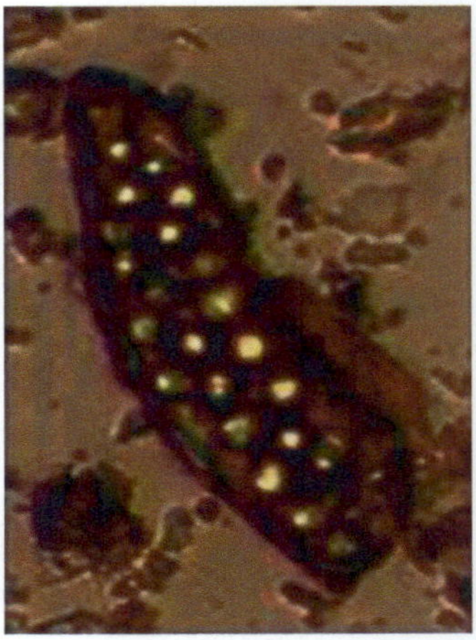

FIG. 7.5.9 Simple starch grains embedded in parenchymatous cells in *Saraca asoca* bark powder.

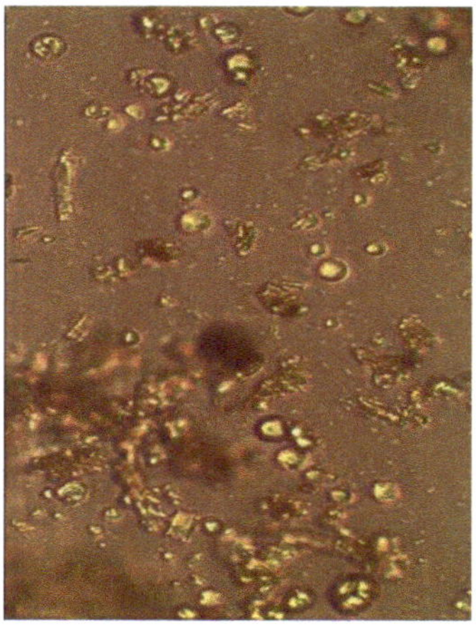

FIG. 7.5.10 Starch grains present in *Polyalthia longifolia* bark powder.

References

[1] Government of India, Department of AYUSH, Ayurvedic Pharmacopoeia of India. Part I, vol. I, Ministry of Health and Welfare, Ghaziabad, 1999–2017, p. 17.

[2] A.K. Gupta, N. Tandon, M. Sharma, Quality Standards of Indian Medicinal Plants, vol. 2, Medicinal Plant Unit, ICMR, New Delhi, India, 2008, p. 201.

[3] M.G. Chauhan, A.P.G. Pillai, Microscopic Profile of Powdered Drugs Used in Indian System of Medicine, Bark Drugs, vol. 1, Institute of Ayurvedic Medicinal Plant Sciences, Jamnagar, Gujarat, India, 2005, p. 18.

Chapter 7.6

Terminalia arjuna — Terminalia bellirica, Terminalia calamansi, Terminalia chebula

Terminalia is a genus of large trees of the flowering plant family Combretaceae, comprising around 100 species distributed in tropical regions of the world. This genus gets its name from Latin *terminus*, referring to the fact that the leaves appear at the very tips of the shoots. *T. arjuna* species has potent cardiovascular properties and commonly substituted with *T. bellirica, T. calamansi* and *T. chebula*.

Terminalia arjuna (Roxb. ex DC.) Wight & Arn.	Combretaceae
Terminalia bellirica (Gaertn.) Roxb	Combretaceae
Terminalia calamansi Rolfe	Combretaceae
Terminalia chebula Retz.	Combretaceae

Common vernacular names: Assamese – Arjun; Bengali – Arjuna; Gujarati – Sadad, Arjuna, Sajada; Hindi – Arjuna; Kannada – Matti, Bilimatti, Neermatti, Mathichakke, Kudare Kivimase; Malayalam – Nirmasuthu, Vellamaruthi, Kellemasuthu, Mattimora, Torematti; Marathi – Arjuna, Sadada; Oriya – Arjuna; Punjabi – Arjon; Tamil – Marudam; Telugu – Maddi [1].

Major chemical constituents: Cardiac glycosides [2].

Traditional uses: Cardiotonic, Astringent [2].

Macroscopic description: The morphological characters of *T. arjuna, T. bellirica, T. calamansi, T. chebula* are illustrated in Table 7.6.1 and Figs. 7.6.1–7.6.4 respectively.

7.6.1 Powder microscopic description

T. arjuna bark powder is pinkish brown in colour showing abundant cluster and rosette crystals of calcium oxalate of various shapes and sizes, scattered as such; Fragments of thin walled, long fibres; Fragments of radially and tangentially longitudinally cut medullary rays and abundant starch grains and cork cells scattered throughout the field [3].

Powdered Crude Drug Microscopy of Leaves and Barks. https://doi.org/10.1016/B978-0-12-818092-1.00050-6

TABLE 7.6.1 Morphological characters of four species of *Terminalia*.

S. No.	Features	*T. arjuna*	*T. bellirica*	*T. calamansi*	*T. chebula*
1	Shape	Flat, curved and channeled	Thick and flat pieces	Flat or recurved	Thick and recurved
2	Outer bark	Smooth and grey	Grey whitish brown, rough and somewhat warty	Light brown and slightly rough	Greyish brown, rough and warty
3	Inner bark	Fibrous and pinkish	Light brown, fibrous and longitudinally ridged	Reddish brown with striations	Dark brown with longitudinal fine folds
4	Fracture	Inner short and outer laminated	Fibrous and hard	Short and fibrous	Hard and coarse
5	Odour	Characteristic	Not characteristic	Not characteristic	Not characteristic

FIG. 7.6.1 *Terminalia arjuna* bark.

FIG. 7.6.2 *Terminalia bellirica* bark.

FIG. 7.6.3 *Terminalia calamansi* bark.

T. bellirica bark powder is greyish brown in colour showing few cluster, rosette and prismatic crystals of calcium oxalate of various shapes and sizes, scattered as such; Fragments of septate thick or thin walled, fibres; Fragments of radially and tangentially longitudinally cut medullary rays; Dark brown masses containing tannin cells and abundant starch grains and cork cells scattered throughout the field.

FIG. 7.6.4 *Terminalia chebula* bark.

T. chebula bark powder is light reddish brown in colour showing few cluster, rosette and prismatic crystals of calcium oxalate of various shapes and sizes, scattered as such; fragments of septate thin walled, bifurcating fibres; Fragments of radially and tangentially longitudinally cut medullary rays and abundant starch grains and cork cells scattered throughout the field.

T. calamansi bark powder is dark brown in colour showing few cluster and rosette crystals of calcium oxalate of various shapes and sizes, scattered as such; Fragments of abundant septate, thin walled fibres present in groups; Fragments of radially and tangentially longitudinally cut medullary rays and abundant starch grains and cork cells scattered throughout the field.

Based upon the powder microscopic features, the bark powders of *T. arjuna, T. bellirica, T. calamansi, T. chebula* can be distinguished as shown in Table 7.6.2 and Figs. 7.6.5–7.6.15. Bark powders of *Terminalia* differed in terms of abundance of rosette crystals. *T. arjuna* bark powder was richest in rosette crystals and compared to other three species whereas prismatic crystals and dark brown cells containing tannins were seen in *T. bellirica* bark powder. The fibres were abundantly present in *T. bellirica* and *T. calamansi*. Fibres were thick walled and sepatate in *T. bellirica* whereas they were finely striated and thin walled in other three species.

TABLE 7.6.2 Specific comparative feature of powdered bark powders of four species of *Terminalia.*

S. No.	Features	*Terminalia arjuna*	*T. bellirica*	*T. chebula*	*T. calamansi*
1	**Common name**	Arjuna	Bahera	Harad	Kako
2	**Calcium oxalate crystals**	Rosette	Rosette	Rosette	Rosette
		Cluster	Cluster	Cluster	Cluster
			Prismatic	Prismatic	
3	**Fibres**				
	Occurrence	Few	Abundant	Few	Abundant
		Isolated	Isolated	Isolated	Group
	Walls	Smoothly striated	Septate	Thin walled	Thin walled
		Thin walled	Thick walled	Bifurcating ends	Crystal fibres
				Striations	Irregular swellings
					Faintly striated
					Septate
	Shape	Very long	Very long	Very long	Very long
		Irregular margin	Pointed apex		Bent
		Rounded			
	Lumen	Narrow	Narrow	Narrow	Narrow
		Uneven			Uneven
4	**Masses**	NA	Dark brown cells containing tannins	NA	NA

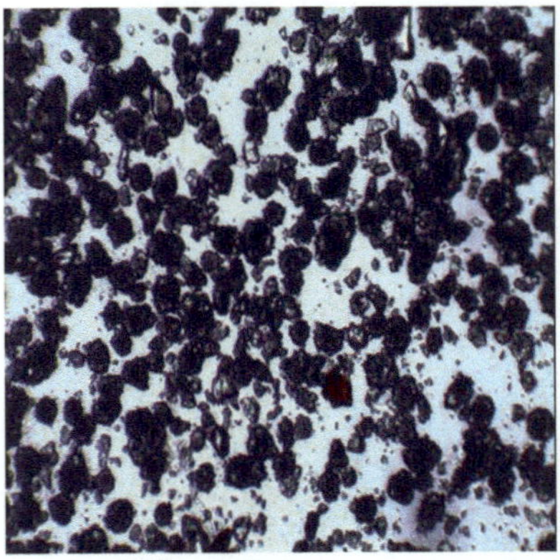

FIG. 7.6.5 Rosette and cluster crystals from *Terminalia arjuna* bark powder.

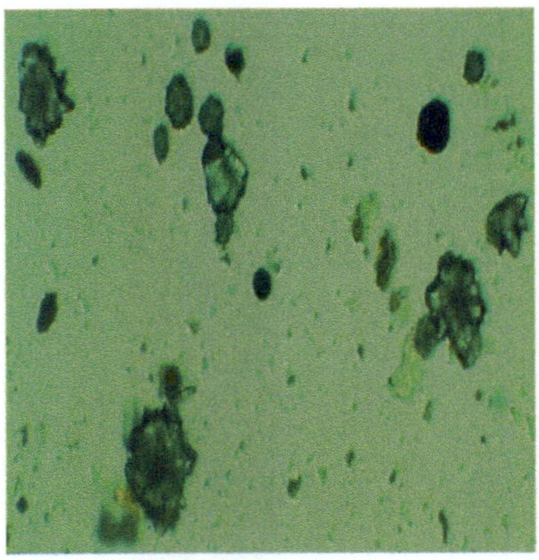

FIG. 7.6.6 Rosette and cluster crystals from *Terminalia bellirica* bark powder.

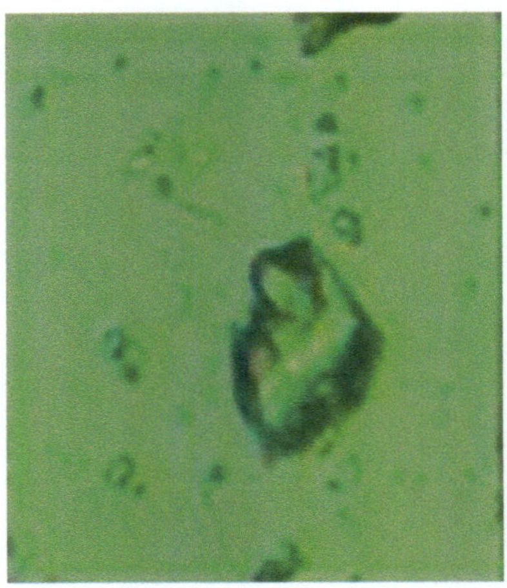

FIG. 7.6.7 Prismatic crystals from *Terminalia bellirica* bark powder.

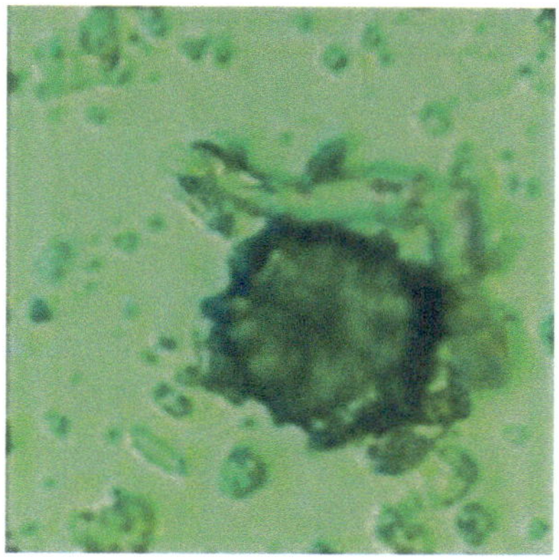

FIG. 7.6.8 Cluster crystals from *Terminalia chebula* bark powder.

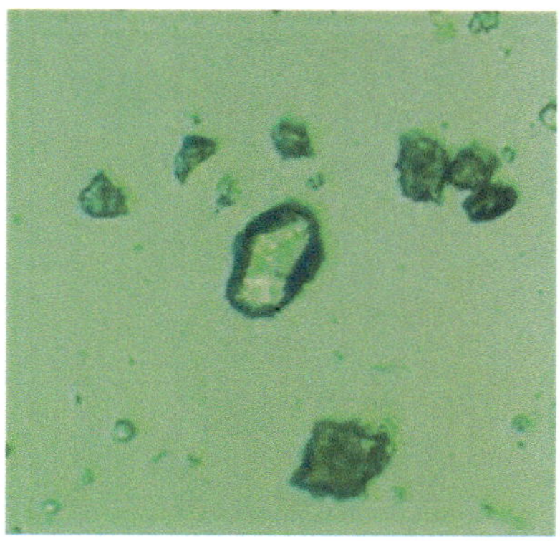

FIG. 7.6.9 Prismatic crystals from *Terminalia chebula* bark powder.

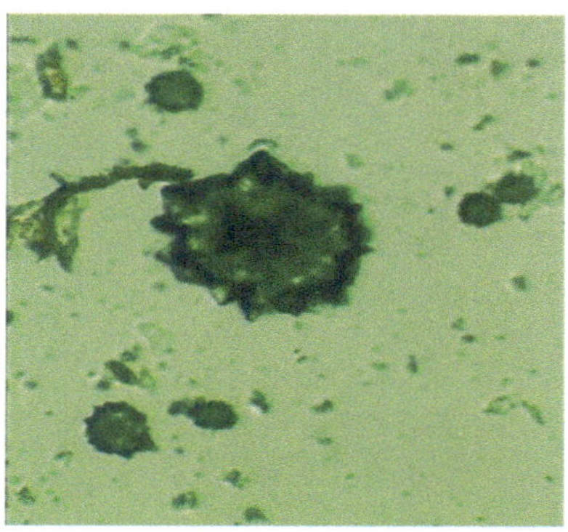

FIG. 7.6.10 Rosette and cluster crystals from *Terminalia calamansi* bark powder.

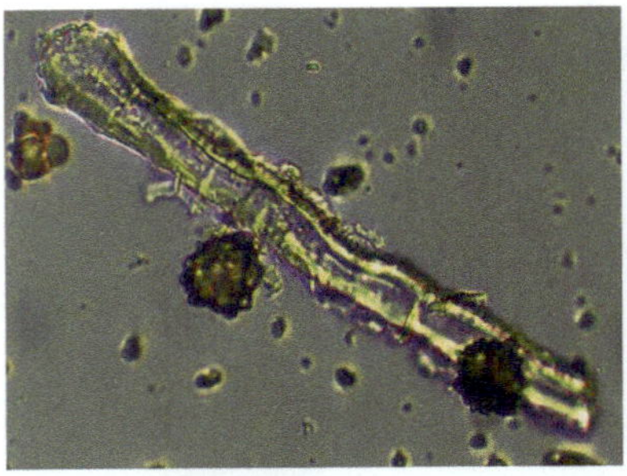

FIG. 7.6.11 Fibre from *Terminalia arjuna* bark powder.

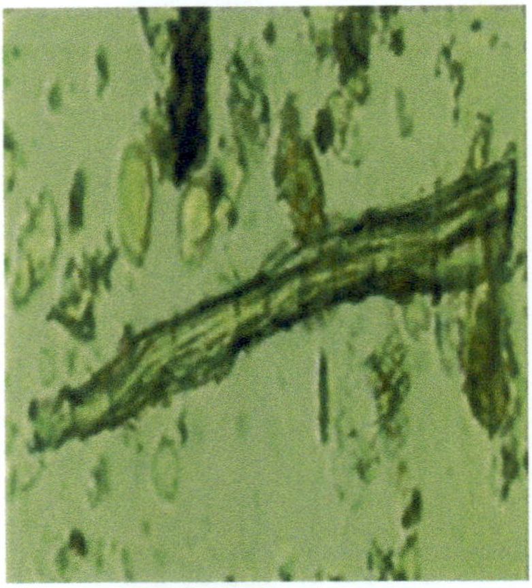

FIG. 7.6.12 Fibre from *Terminalia bellirica* bark powder.

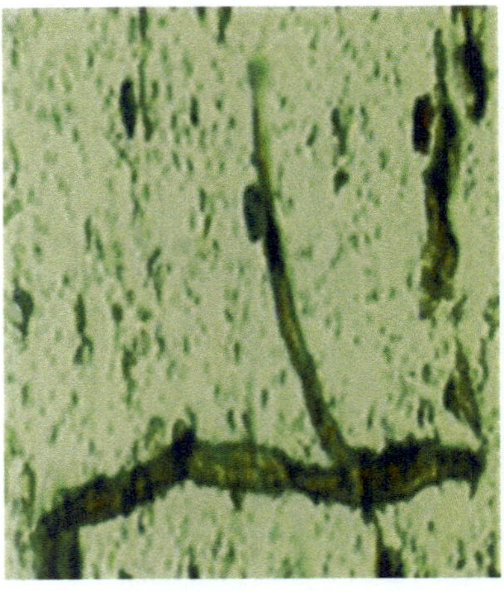

FIG. 7.6.13 Fibre from *Terminalia chebula* bark powder.

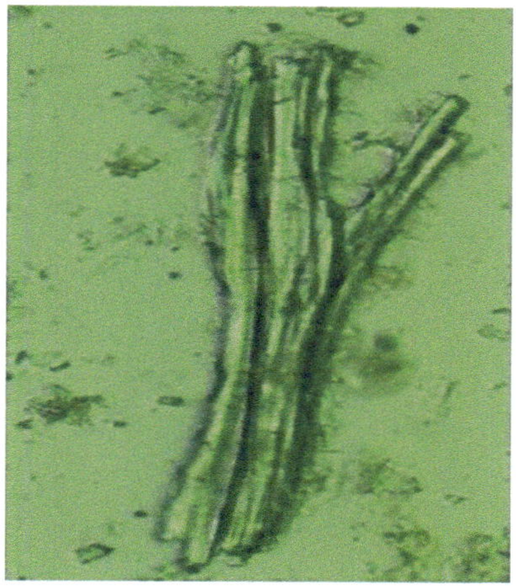

FIG. 7.6.14 Fibre from *Terminalia calamansi* bark powder.

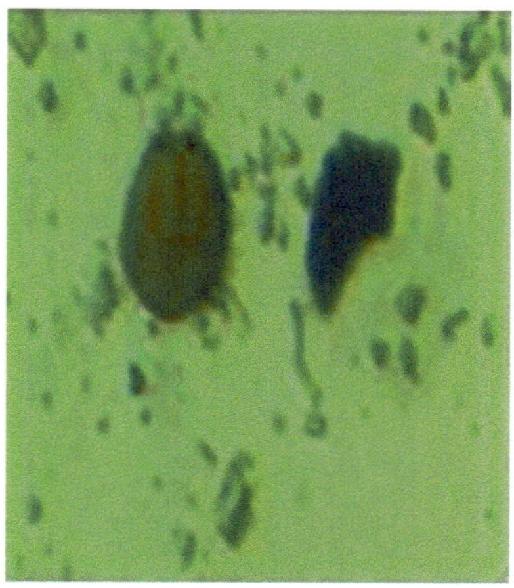

FIG. 7.6.15 Dark brown cell from *Terminalia bellirica* bark powder.

References

[1] Government of India, Department of AYUSH, Ayurvedic Pharmacopoeia of India. Part I, vol. II, Ministry of Health and Welfare, Ghaziabad, 1999–2017, p. 18.

[2] A.K. Gupta, N. Tandon, M. Sharma, Quality Standards of Indian Medicinal Plants, vol. 2, Medicinal Plant Unit, ICMR, New Delhi, India, 2008, p. 244.

[3] M.G. Chauhan, A.P.G. Pillai, Microscopic Profile of Powdered Drugs Used in Indian System of Medicine, Bark Drugs, vol. 1, Institute of Ayurvedic Medicinal Plant Sciences, Jamnagar, Gujarat, India, 2005, p. 12.

Leaf drugs

Chapter 7.7

Adhatoda vasica – Ailanthus excelsa and *Adenanthera pavonina*

Adhatoda vasica is a medicinal plant native to Asia, widely used in Siddha Medicine, Ayurvedic, homeopathy and Unani systems of medicine. It is commonly adulterated with the leaves of *Adenanthera pavonina and Ailanthus excelsa.*

Adhatoda vasica Nees	Acanthaceae
A. excelsa Roxb.	Simaroubaceae
A. pavonina L.	Fabaceae

Common vernacular names: Assamese – Titabahak, Bahak, Vachaka; Bengali – Baksa, Vasaka; English – Vasaka; Gujarati – Aduso, Ardusi, Adulso; Hindi – Aduss, Arusa; Kannada – Adsale, Adusoge, Atarusha, Adsole, Adasale; Kashmiri – Vasa; Malayalam – Attalatakam, Atalotakam; Marathi – Vasa, Adulsa; Oriya – Basanga; Punjabi – Bhekar, Vansa, Arusa; Tamil – Vasambu, Adathodai; Telugu – Addasaramu [1].

Major chemical constituents: Vasicine – Quinazoline alkaloid [2].

Traditional uses: Dyspnoea and cough [2].

Macroscopic description: The morphological characters of *Adhatoda vasica, A. pavonina* and *A. excelsa* are illustrated in Table 7.7.1 and Figs. 7.7.1–7.7.3 respectively.

7.7.1 Powder microscopic description

Adhatoda vasica leaf powder is dark green in colour showing plenty of simple, covering multicellular, uniseriate and sessile glandular trichomes of various sizes and shapes scattered as such, in broken fragments or attached to the cells of epidermis; Simple trichomes are two to five celled long, straight or bent with acute, blunt or thread like terminals. Majority of glandular trichomes are sessile with a short stalk and bicellular head; Fragments of upper and lower epidermis in surface view embedded with diacytic stomata, cicatrix, with straight anticlinical walls and few prismatic crystals of calcium oxalate [3].

Powdered Crude Drug Microscopy of Leaves and Barks. https://doi.org/10.1016/B978-0-12-818092-1.00051-8

TABLE 7.7.1 **Morphological characters of four species of *Adhatoda vasica, Ailanthus excelsa* and *Adenanthera pavonina*.**

S. No.	Features	*Adhatoda vasica*	*Ailanthus excelsa*	*Adenanthera pavonina*
1	Shape	Ovate to ovate-lanceolate	Oval or lance shaped	Oval to oblong
2	Apex	Acuminate	Pointed	Obtuse
3	Base	Symmetrical and tapering	Unequal	Acute, slightly asymmetrical
4	Margin	Slightly crenate, wavy to entire	Coarsely toothed	Entire
5	Venation	Pinnate	Pinnate	Pinnate
6	Petiole	Cylindrical, slightly grooved on the upper surface	Very long	Short
7	Colour	Dark green on the upper surface and pale green and somewhat glabrous on the lower surface	Light green	Dark green on the upper surface and paler on the lower surface
8	Odour	Not characteristic	Characteristic	Characteristic

FIG. 7.7.1 *Adhatoda vasica* leaf.

FIG. 7.7.2 *Ailanthus excelsa* leaf.

FIG. 7.7.3 *Adenanthera pavonina* leaf.

A. excelsa leaf powder is dark green in colour showing plenty of unicellular, straight or bent thick walled trichomes; Rosette crystals of calcium oxalate and fragments of upper and lower epidermis with straight and beaded walls devoid of stomata.

A. pavonina leaf powder is light green in colour showing fragments of straight and beaded epidermal walls devoid of stomata; Few prismatic crystals of calcium oxalate and trichomes are found to be absent in the powder.

TABLE 7.7.2 Specific comparative feature of powdered bark powders of *Adthatoda vasica, Ailanthus excelsa* and *Adenanthera pavonina.*

S. No.	Features	*Adhatoda vasica*	*Ailanthus excelsa*	*Adenanthera pavonina*
1	**Common name**	Arusa, Vasaka	Maharuk	Red lucky seed
2	**Epidermis**	Straight	Straight	Straight
		Anticlinical walls	Beaded	Beaded
3	**Stomata**	Diacytic	Devoid of stomata	Devoid of stomata
4	**Trichomes**	Abundant	Abundant	NA
		Simple	Simple	
		Covering	Covering	
		Sessile	Straight	
		Glandular	Unicellular	
		Unicellular	Thick walled	
		Multicellular	Bent	
		Uniseriate		
		Short stalk unicellular		
		4-Celled head		
		Cicatrix		
		Globular head		
5	**Calcium oxalate crystals**	Prismatic	Prismatic	Rosette

Based upon the powder microscopic features, the leaf powders of *Adthatoda vasica, A. excelsa* and *A. pavonina* can be distinguished as shown in Table 7.7.2 and Figs. 7.7.4–7.7.11. Leaf powder of *A. vasica* showed straight epidermal walls embedded with diacytic stomata whereas *A. excelsa* and *A. pavonina* epidermal walls are beaded and devoid of stomata. The trichomes are abundantly present in *A. vasica* ranging from simple, covering to sessile, glandular whereas simple, covering, straight trichomes are present in *A. excelsa* and trichomes are not observed in *A. pavonina*. Rosette crystals of calcium oxalate are observed in *A. pavonina* as compared to prismatic crystals in *A. vasica* and *A. excelsa*.

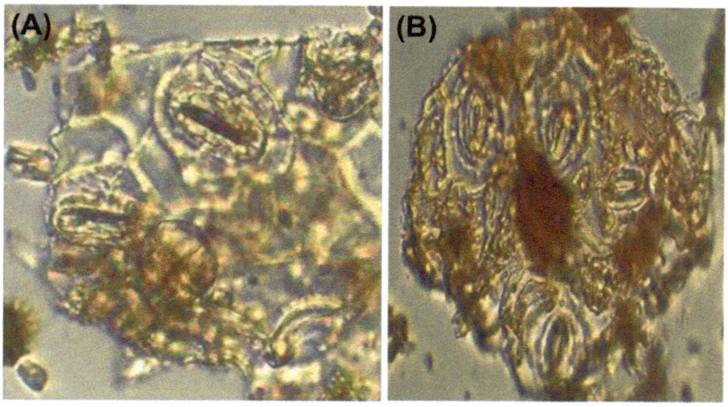

FIG. 7.7.4 (A–B) – Diacytic stomata present in *Adhatoda vasica* leaf powder.

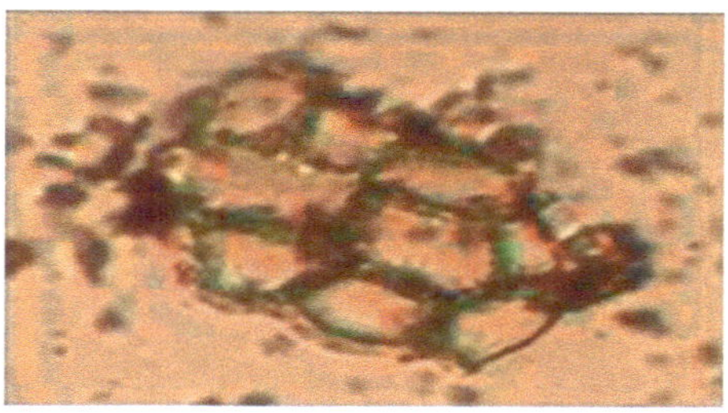

FIG. 7.7.5 Epidermis present in *Ailanthus excelsa* leaf powder.

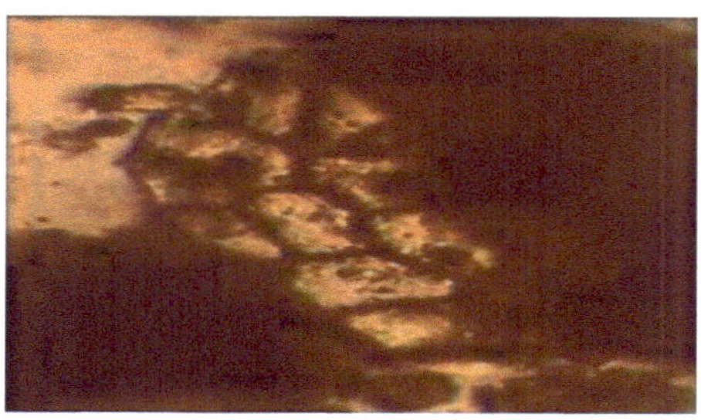

FIG. 7.7.6 Epidermis present in *Adenanthera pavonina* leaf powder.

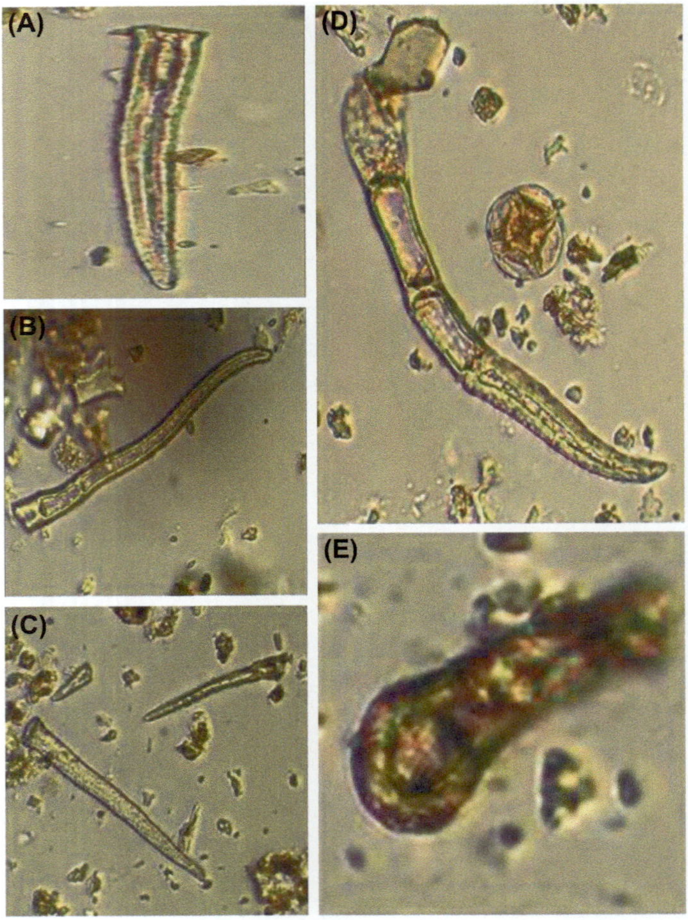

FIG. 7.7.7 (A–E) – Various types of trichomes present in *Adhatoda vasica* leaf powder.

FIG. 7.7.8 Simple covering trichomes present in *Ailanthus excelsa* leaf powder.

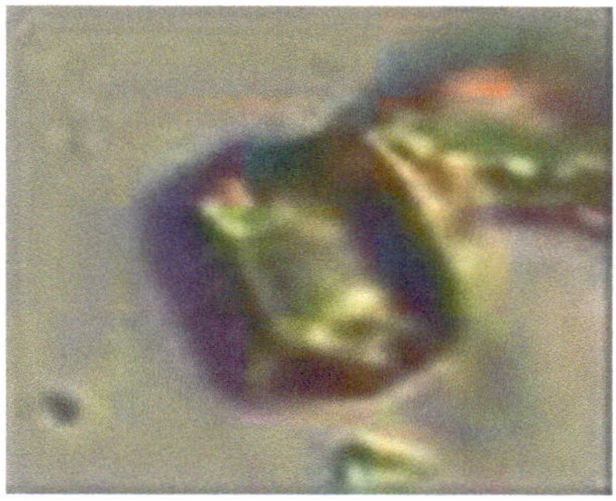

FIG. 7.7.9 Prismatic crystals present in *Adhatoda vasica* leaf powder.

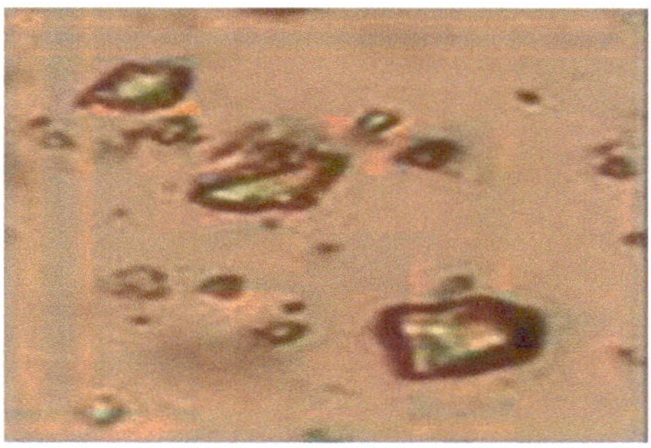

FIG. 7.7.10 Prismatic crystals present in *Ailanthus excelsa* leaf powder.

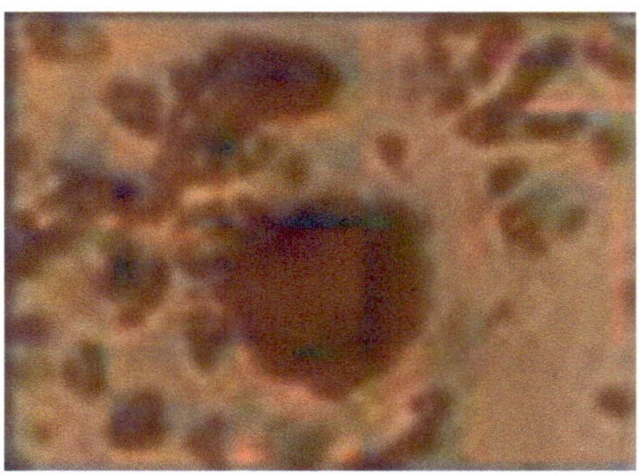

FIG. 7.7.11 Rosette crystals present in *Adenanthera pavonina* leaf powder.

References

[1] Government of India, Department of AYUSH, Ayurvedic Pharmacopoeia of India. Part I, vol. I, Ministry of Health and Welfare, Ghaziabad, 1999–2017, p. 122.

[2] A.K. Gupta, N. Tandon, M. Sharma, Quality Standards of Indian Medicinal Plants, vol. 11, Medicinal Plant Unit, ICMR, New Delhi, India, 2008, p. 27.

[3] M.G. Chauhan, A.P.G. Pillai, Microscopic Profile of Powdered Drugs Used in Indian System of Medicine, Leaf Drugs, vol. 2, Institute of Ayurvedic Medicinal Plant Sciences, Jamnagar, Gujarat, India, 2007, p. 168.

Chapter 7.8

Azadirachta indica and *Melia azadirach*

Azadirachta indica, commonly known as neem or Indian lilac, is a tree in the mahogany family Meliaceae. It is one of two species in the genus *Azadirachta*, and is native to the Indian subcontinent, i.e., India, Nepal, Pakistan, Bangladesh, Sri Lanka, and Maldives. According to the literature, Neem and Persian lilac have very often been confused with each other in spite of the fact that they are morphologically different.

A. indica A.Juss	Meliaceae
Melia azadirach L.	Meliaceae

Common vernacular names: Assamese – Mahanim; Bengali – Nim, Nimgach; English – Margosa Tree; Gujarati – Limba, Limbado, Limado, Kohumba; Hindi – Nim, Nimba; Kannada – Nimba, Bevu, Oilevevu, Kahi-bevu, Bevinama; Malayalam – Veppu, Aryaveppu, Nimbam, Veppa; Marathi – Balantanimba, Limba, Bakayan, Nim, Kadunimb; Oriya – Nimba; Punjabi – Nimba, Bakan, Nim; Tamil – Vemmu, Veppu, Arulundi, Veppan; Telugu – Vemu, Vepa [1].

Major chemical constituents: Quercetin, β-sitosterol, nimocinolide, iso-nimocinolide [2].

Traditional uses: Anthelmentic, dermatopathies, fever and anorexia [2].

Macroscopic description: The morphological characters of *A. indica* and *M. azadirach* are illustrated in Table 7.8.1 and Figs. 7.8.1 and 7.8.2 respectively.

7.8.1 Powder microscopic description

A. indica leaf powder is yellowish green in colour showing epidermis with straight anticlinical walls embedded with anomocytic and actinocytic stomata; Simple, unicellular, thick walled, curved trichomes scattered as such; Prismatic crystals of calcium oxalate and crystal fibres embedded with idioblasts [3].

M. azadirach leaf powder is yellowish green in colour showing epidermis with wavy walls embedded with anomocytic stomata; Glandular and stellate trichomes scattered as such and needles of calcium oxalate spreaded throughout the field.

Powdered Crude Drug Microscopy of Leaves and Barks. https://doi.org/10.1016/B978-0-12-818092-1.00052-X

TABLE 7.8.1 Morphological characters of *Azadirachta indica* and *Melia azadirach*.

S. No.	Features	*Azadirachta indica*	*Melia azadirach*
1	Shape	Lanceolate	Ovate to oblong lanceolate
2	Apex	Acute	Acuminate
3	Base	Oblique	Slightly unequilateral, acute to rounded
4	Margin	Serrate	Entire to variously serrate
5	Venation	Reticulate	Pinnate
6	Petiole	Shortly petioluled	Long and cylindrical
7	Colour	Slightly yellowish green and glabrous	Dark green
8	Odour	Indistinct	Pungent odour when crushed

FIG. 7.8.1 *Azadirachta indica* leaf.

FIG. 7.8.2 *Melia azadirach* leaf.

Based upon the powder microscopic features, the leaf powders of *A. indica* and *M. azadirach* can be distinguished as shown in Table 7.8.2 and Figs. 7.8.3–7.8.9. The epidermal walls of *A. indica* are straight embedded with anomocytic and actinocytic stomata as compared to wavy epidermal walls in *M. azadirach.* The trichomes of *A. indica* are simple, covering and warty whereas stellate trichomes are present in *M. azadirach.* Prismatic crystals are present in *A. indica* as compared to needle shaped crystals in *M. azadirach.*

TABLE 7.8.2 Specific comparative feature of powdered leaf powders of *Azadirachta indica* and *Melia azadirach*.

S. No.	Features	*Azadirachta indica*	*Melia azadirach*
1	**Common name**	Neem	Bakayana
2	**Epidermis**	Straight	Wavy
		Anticlinical	
3	**Trichomes**	Simple	Glandular
		Covering	Stellate
		Warty	
4	**Stomata**	Anomocytic	Anomocytic
		Actinocytic	
5	**Calcium oxalate crystals**	Prismatic	Needles

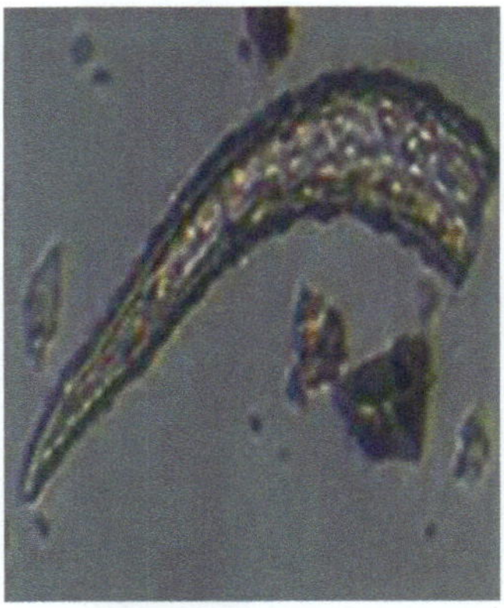

FIG. 7.8.3 Simple trichome present in *Azadirachta indica* leaf powder.

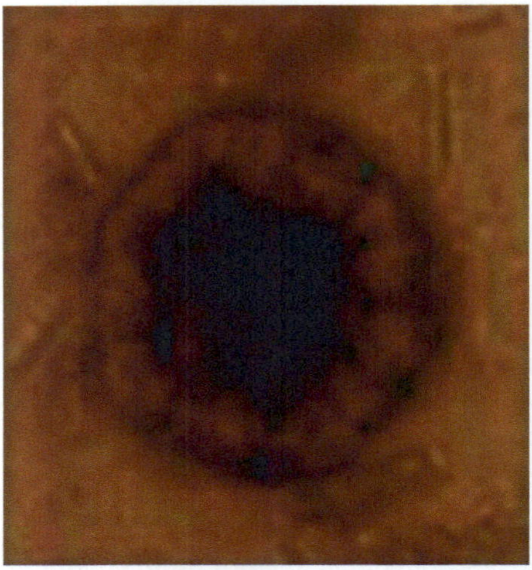

FIG. 7.8.4 Glandular trichome present in *Melia azadirach* leaf powder.

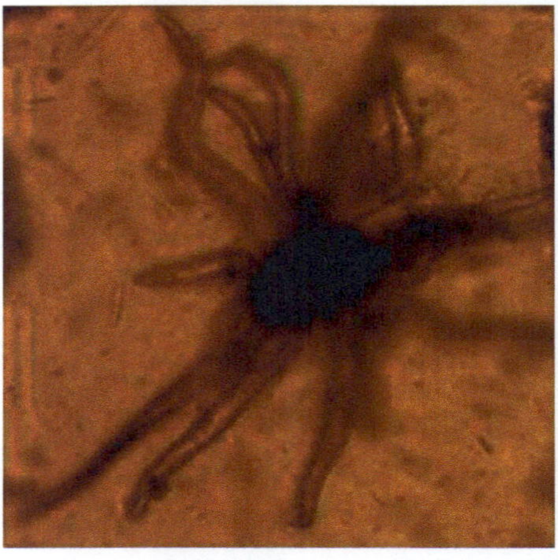

FIG. 7.8.5 Stellate trichome present in *Melia azadirach* leaf powder.

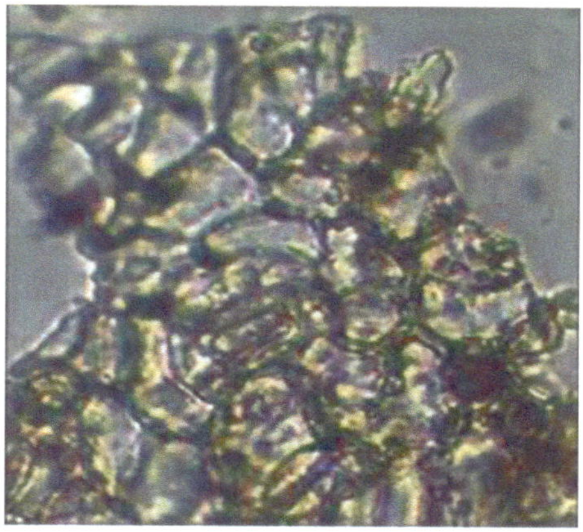

FIG. 7.8.6 Actinocytic stomata present in *Azadirachta indica* leaf powder.

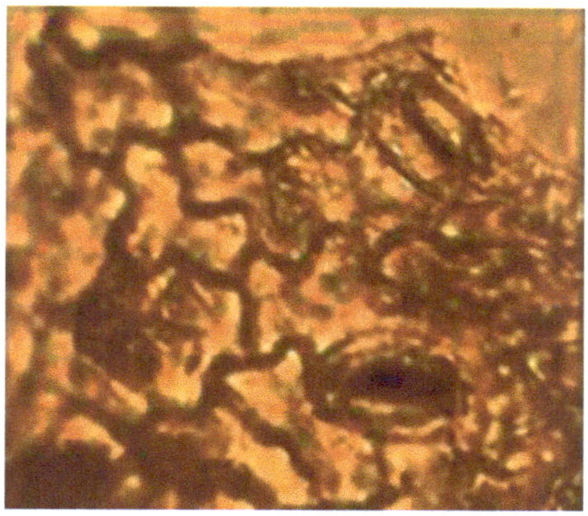

FIG. 7.8.7 Anomocytic stomata present in *Melia azadirach* leaf powder.

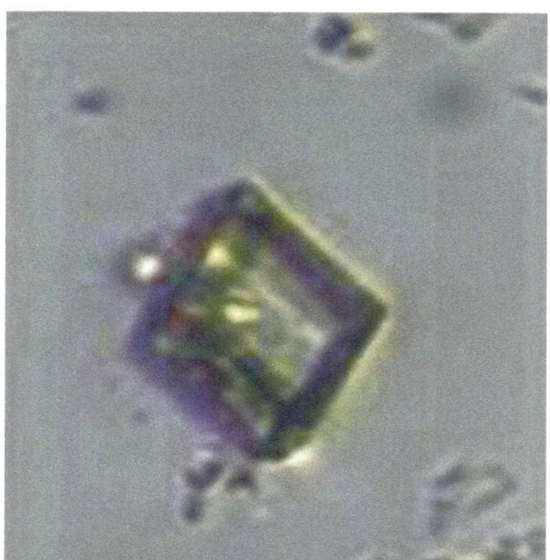

FIG. 7.8.8 Prismatic crystals present in *Azadirachta indica* leaf powder.

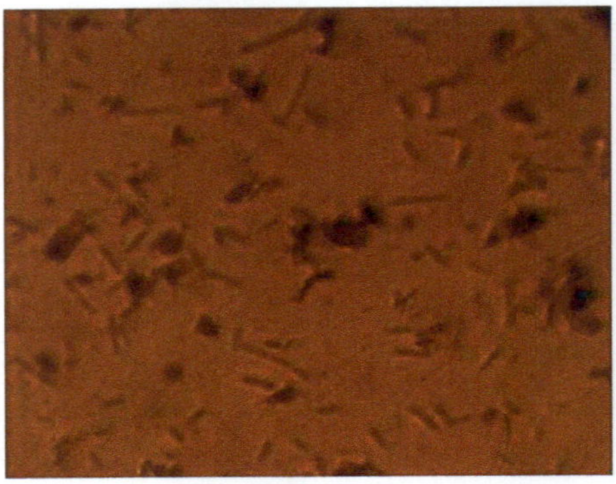

FIG. 7.8.9 Needle shaped crystals present in *Melia azadirach* leaf powder.

References

[1] Government of India, Department of AYUSH, Ayurvedic Pharmacopoeia of India. Part I, vol. II, Ministry of Health and Welfare, Ghaziabad, 1999–2017, p. 132.

[2] A.K. Gupta, N. Tandon, M. Sharma, Quality Standards of Indian Medicinal Plants, vol. 11, Medicinal Plant Unit, ICMR, New Delhi, India, 2008, p. 83.

[3] M.G. Chauhan, A.P.G. Pillai, Microscopic Profile of Powdered Drugs Used in Indian System of Medicine, Leaf Drugs, vol. 2, Institute of Ayurvedic Medicinal Plant Sciences, Jamnagar, Gujarat, India, 2007, p. 98.

Chapter 7.9

Cassia angustifolia — *Cassia auriculata* and *Ailanthus glandulosa*

Senna is a commonly found herbal plant traditionally used by Arabian physicians primarily as a cathartic. Senna leaves have been commonly adulterated by *Cassia auriculata* and *Ailanthus glandulosa* leaves.

Cassia angustifolia M. Vahl	Leguminosae
C. auriculata L.	Fabaceae
A. glandulosa Desf.	Simaroubaceae

Common vernacular names: Assamese — Sonamukhi; Bengali — Svamamukhi, Sonapata; English — Indian Senna, Tinnevelly Senna; Gujarati — Mindhiaval, Sonamukhi; Hindi — Sanaya, Hindisana; Kannada — Nelavarika, Sonamukhi, Nelaavare, Nelavarike, Nela Aanriake; Kashmiri — Sna; Malayalam — Sunnamukhi, Nilavaka, Chinnukki, Adapatiyan; Marathi — Sonamukhi; Oriya — Sunamukhi; Punjabi — Sannamakhi, Sanapati, Sarnapatta; Tamil — Nilapponnai, Avarai; Telugu — Sunamukhi [1].

Major chemical constituents: Anthraquinone glycosides [1].

Traditional uses: Laxative in constipation [1].

Macroscopic description: The morphological characters of *C. angustifolia, C. auriculata* and *A. glandulosa* are illustrated in Table 7.9.1 and Figs. 7.9.1—7.9.3 respectively.

7.9.1 Powder microscopic description

C. angustifolia is yellowish green in colour showing unicellular thick walled, warty, straight or bent trichomes scattered as such throughout; Clusters and prismatic crystals of Calcium oxalate; Few simple spherical starch grains; Epidermis shows straight anticlinical walls embedded with paracytic and anomocytic stomata [2].

C. auriculata is light green in colour showing unicellular thick walled, simple covering trichomes scattered as such throughout; Few simple spherical starch grains and epidermis shows straight anticlinical walls embedded with few anomocytic stomata.

Powdered Crude Drug Microscopy of Leaves and Barks. https://doi.org/10.1016/B978-0-12-818092-1.00053-1

TABLE 7.9.1 Morphological characters of *Cassia angustifolia, Cassia auriculata* and *Ailanthus glandulosa.*

S. No.	Features	*Cassia angustifolia*	*Cassia auriculata*	*Ailanthus glandulosa*
1	Shape	Elongated leaflets and lanceolate	Ovate to lanceolate	Ovate-lanceolate
2	Apex	Flat apex, acute with a sharp spine	Obtuse	Acute to acuminate
3	Base	Slightly asymmetric at base	Rounded	Truncate
4	Margin	Entire	Entire	Entire
5	Venation	Pinnate	Pinnate	Alternate
6	Petiole	Glandless	Short	Short and cylindrical
7	Colour	Pale yellowish green and smooth on both surfaces	Dark green to light green	Dark green on the upper and pale green on the lower
8	Odour	Faint but distinctive	Characteristic	Unpleasant

FIG. 7.9.1 *Cassia angustifolia* leaf.

FIG. 7.9.2 *Cassia auriculata* leaf.

FIG. 7.9.3 *Ailanthus glandulosa* leaf.

A. glandulosa is light green in colour showing unicellular, simple covering, very long trichomes and epidermis shows straight anticlinical walls with few diacytic stomata.

Based upon the powder microscopic features, the leaf powders of *C. angustifolia, C. auriculata* and *A. glandulosa* can be distinguished as shown in Table 7.9.2 and Figs. 7.9.4–7.9.11. Leaf powder of *C. angustifolia, C.*

TABLE 7.9.2 Specific comparative feature of powdered leaf powders of *Cassia angustifolia, Cassia auriculata* and *Ailanthus glandulosa.*

S. No.	Features	*Cassia angustifolia*	*Cassia auriculata*	*Ailanthus glandulosa*
1	**Common name**	Tinnevelly senna	Palthe senna	Foul smelling tree
2	**Stomata**	Paracytic	Anomocytic	Diacytic
		Anomocytic		
3	**Trichomes**	Unicellular	Unicellular	Abundant
		Thick walled	Thick walled	Unicellular
		Simple	Straight	Simple
		Covering	Bent	Covering
		Straight	Covering	Uniseriate
		Bent	Long	Irregularly bent
		Pointed apex		Pointed apex
		Warty		
4	**Calcium oxalate crystals**	Cluster	NA	NA
		Prismatic		

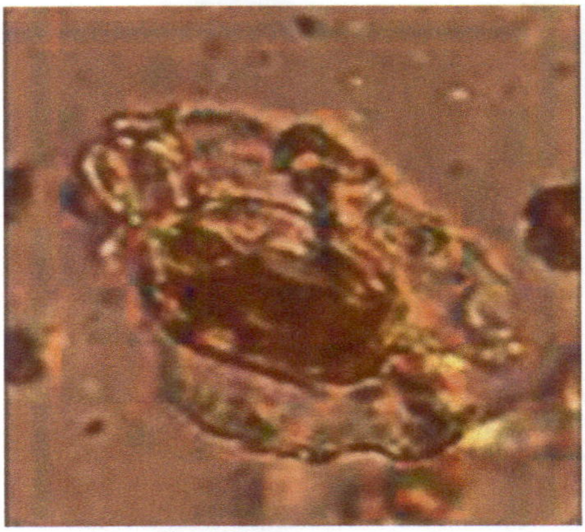

FIG. 7.9.4 Paracytic stomata present in *Cassia angustifolia* leaf powder.

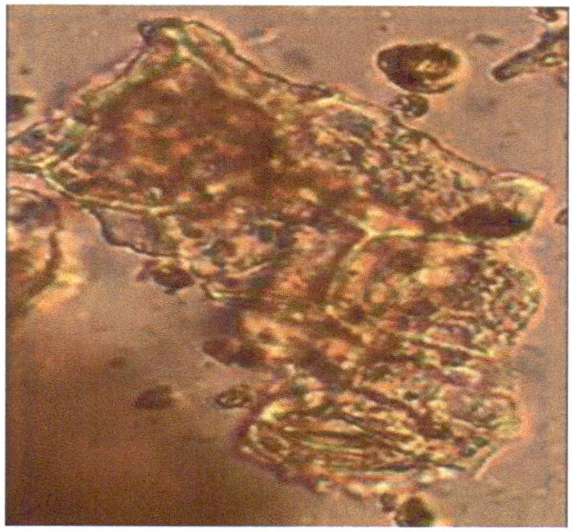

FIG. 7.9.5 Anomocytic stomata present in *Cassia auriculata* leaf powder.

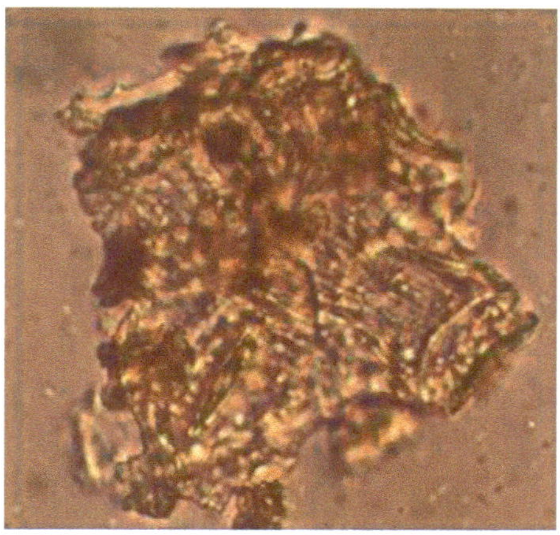

FIG. 7.9.6 Diacytic stomata present in *Ailanthus glandulosa* leaf powder.

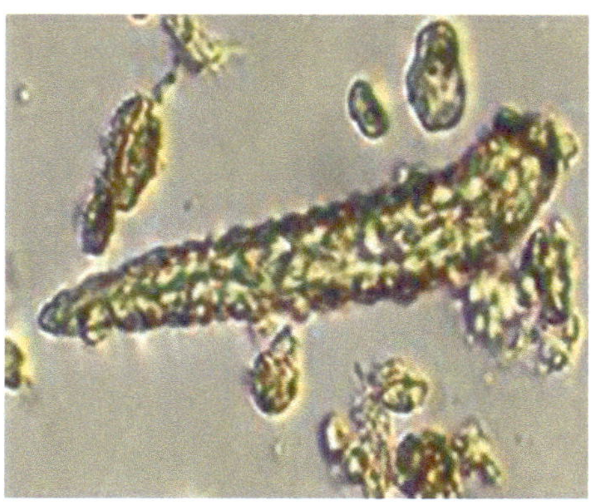

FIG. 7.9.7 Trichome present in *Cassia angustifolia* leaf powder.

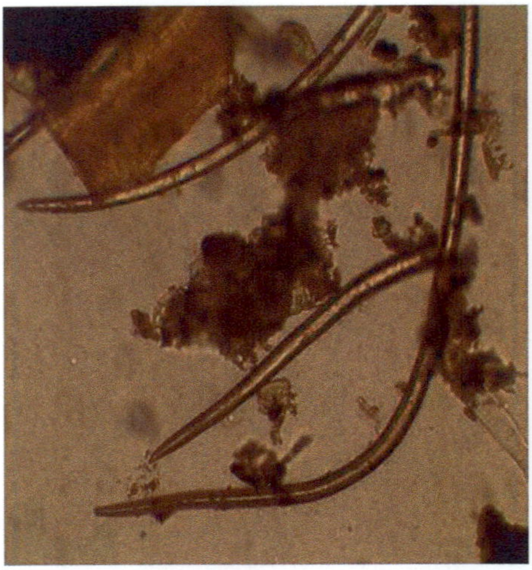

FIG. 7.9.8 Trichome present in *Cassia auriculata* leaf powder.

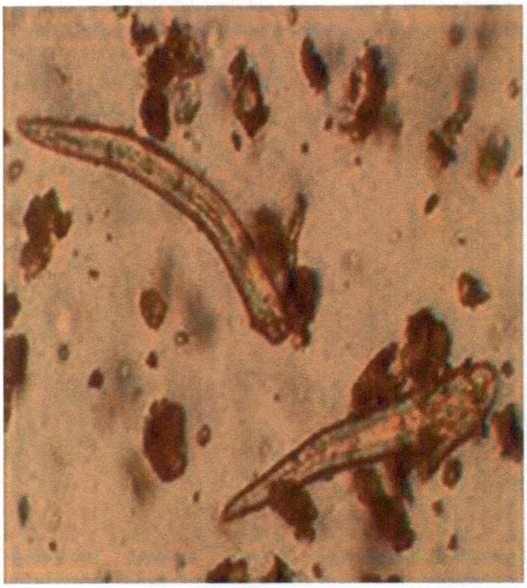

FIG. 7.9.9 Trichome present in *Ailanthus glandulosa* leaf powder.

auriculata and *A. glandulosa* shows paracytic, anomocytic and diacytic stomata respectively. Trichomes of *C. auriculata* are the longest whereas they are warty in case of *C. angustifolia* and simple, unicellular in *A.glandulosa*. Besides, cluster and prismatic crystals of calcium oxalate are observed in *C. angustifolia* leaf powder.

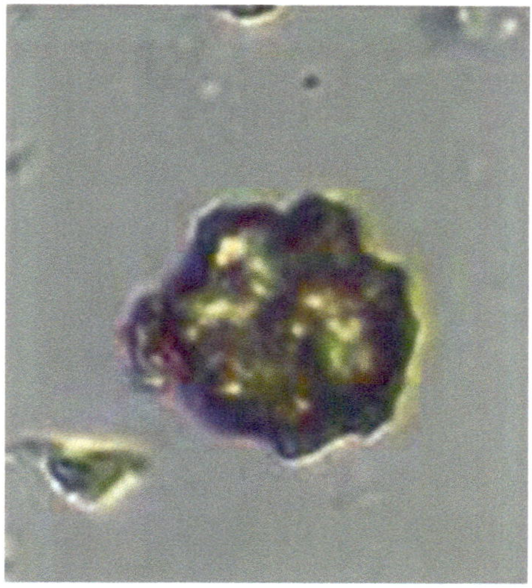

FIG. 7.9.10 Cluster crystal present in *Cassia angustifolia* leaf powder.

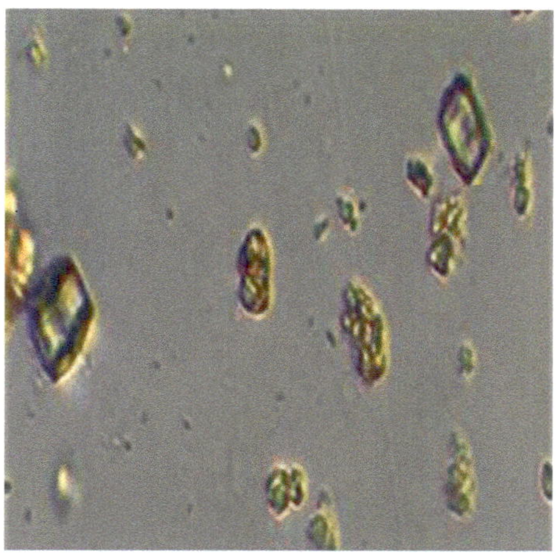

FIG. 7.9.11 Prismatic crystal present in *Cassia angustifolia* leaf powder.

References

[1] Government of India, Department of AYUSH, Ayurvedic Pharmacopoeia of India. Part I, vol. I, Ministry of Health and Welfare, Ghaziabad, 1999–2017, p. 105.

[2] M.G. Chauhan, A.P.G. Pillai, Microscopic Profile of Powdered Drugs Used in Indian System of Medicine, Leaf Drugs, vol. 2, Institute of Ayurvedic Medicinal Plant Sciences, Jamnagar, Gujarat, India, 2007, p. 144.

Chapter 7.10

Indigofera tinctoria and *Indigofera arrecta*

Indigofera is a large genus of over 750 species of flowering plants belonging to the pea family Fabaceae. *Indigofera tinctoria* is used to produce dye indigo whereas *I. arrecta* is an allied species which has been patent to relieve ulcer pain.

I. tinctoria L.	Fabaceae
Indigofera arrecta A.Rich.	Fabaceae

Common vernacular names: Assamese – Nilbam; Bengali – Nil; English – Indigo; Gujarati – Gali, Galiparna; Hindi – Nili; Kannada – Karunili; Malayalam – Neelamar; Marathi – Néel; Oriya – Nili, Nila; Punjabi – Néel; Tamil – Avuri; Telugu – Nili Chettu, Nili [1].

Major chemical constituents: Indican glycoside [1].

Traditional uses: Skin disease, nervous disorders, bronchitis, fever [1].

TABLE 7.10.1 Morphological characters of *Indigofera tinctoria* and *Indigofera arrecta.*

S. No.	Features	*Indigofera tinctoria*	*Indigofera arrecta*
1	Shape	Leaflets oblong to oblanceolate in shape	Ovate
2	Apex	Very short mucronate tip	Acute
3	Base	Rounded	Rounded
4	Margin	Entire	Entire
5	Venation	Pinnate	Pinnate
6	Petiole	Very short circular in shape	Very short
7	Colour	Pale green to greenish black	Dark green on the upper and light green on the lower
8	Odour	No odour	No odour

Powdered Crude Drug Microscopy of Leaves and Barks. https://doi.org/10.1016/B978-0-12-818092-1.00054-3

FIG. 7.10.1 *Indigofera tinctoria* leaf.

FIG. 7.10.2 *Indigofera arrecta* leaf.

Macroscopic description: The morphological characters of *I. tinctoria* and *I. arrecta* are illustrated in Table 7.10.1 and Figs. 7.10.1 and 7.10.2 respectively.

7.10.1 Powder microscopic description

I. tinctoria leaf powder is pale green in colour showing thick walled, highly warty, unicellular and multicellular trichomes with pointed or rounded apex, glandular trichomes are very few, short sessile with unicellular stalk and unicellular head embedded in the epidermis; Anomocytic, anisocytic and few paracytic stomata, their guard cells occasionally being fully or partially collapsed and pitted sclereids [2].

I. arrecta leaf powder is dark green in colour showing simple, unicellular covering trichomes with pointed apex; Prismatic crystals of calcium oxalate; Type of stomata is same as in *I. tinctoria* powder.

Based upon the powder microscopic features, the leaf powders of *I. tinctoria* and *I. arrecta* can be distinguished as shown in Table 7.10.2 and Figs. 7.10.3–7.10.5. Leaf powder of *I. tinctoria* shows multicellular trichome with oval head whereas simple, unicellular trichomes with pointed apex along with few prismatic crystals of calcium oxalate are seen in *I. arrecta* leaf powder.

TABLE 7.10.2 Specific comparative feature of powdered leaf powders of *Indigofera tinctoria* and *Indigofera arrecta*.

S. No.	Features	*Indigofera tinctoria*	*Indigofera arrecta*
1	**Common name**	Nili	Bengal indigo
2	**Trichomes**	Oval head	Simple
		Multicellular	Unicellular
3	**Calcium oxalate crystals**	NA	Prismatic

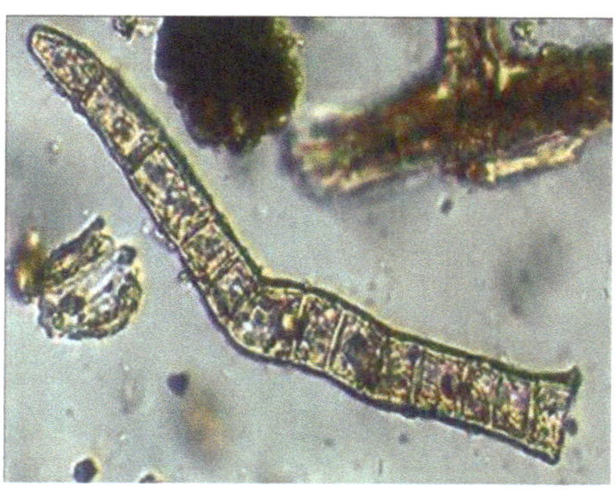

FIG. 7.10.3 Multicellular trichome present in *Indigofera tinctoria* leaf powder.

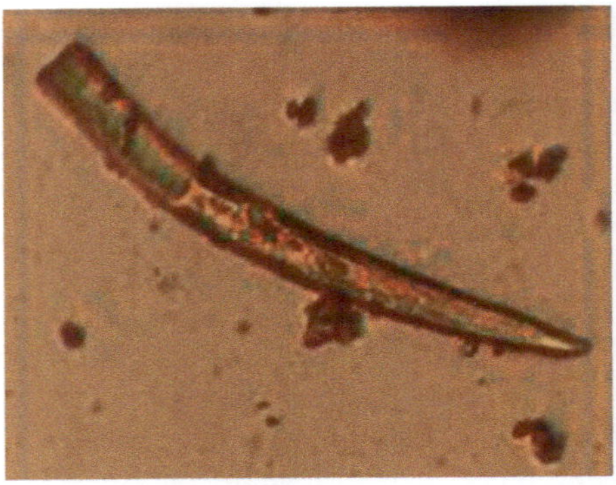

FIG. 7.10.4 Unicellular trichome present in *Indigofera arrecta* leaf powder.

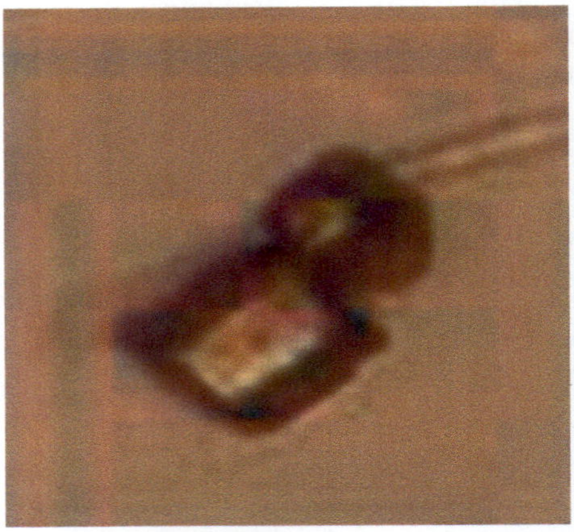

FIG. 7.10.5 Prismatic crystal present in *Indigofera arrecta* leaf powder.

References

[1] Government of India, Department of AYUSH, Ayurvedic Pharmacopoeia of India. Part I, vol. II, Ministry of Health and Welfare, Ghaziabad, 1999–2017, p. 128.

[2] M.G. Chauhan, A.P.G. Pillai, Microscopic Profile of Powdered Drugs Used in Indian System of Medicine, Leaf Drugs, vol. 2, Institute of Ayurvedic Medicinal Plant Sciences, Jamnagar, Gujarat, India, 2007, p. 96.

Chapter 7.11

Phyllanthus fraternus and *Phyllanthus amarus*

The genus *Phyllanthus* consists of more than 1000 species, of which many are used as traditional medicines. *Phyllanthus fraternus* and *P. amarus* are commonly substituted with each other.

P. fraternus G.L.Webster Euphorbiaceae
Phyllanthus amarus Schumach. & Thonn. Phyllanthaceae

Common vernacular names: Assamese – Amlakhi, Amlakhu, Amlaku; Bengali – Amla, Dhatri; English – Emblic Myrobalan; Gujarati – Ambala, Amala; Hindi – Amla, Aonla; Kannada – Nellikayi, Bela nelli, pottade-nollikayi; Kashmiri – Amli, Embali; Malayalam – Nellikka; Marathi – Anvala, Avalkathi; Oriya – Ainla, Anala; Punjabi – Aula, amla; Tamil – Nellikkai, nelli; Telugu – Usirika; Urdu – Amla, Amlaj [1].

Major chemical constituents: Ascorbic acid and gallotannins [1].

Traditional uses: Mild laxative, urinary tract infections [1].

Macroscopic description: The morphological characters of *P. fraternus* and *P. amarus* are illustrated in Table 7.11.1 and Figs. 7.11.1 and 7.11.2 respectively.

TABLE 7.11.1 Morphological characters of *Phyllanthus fraternus* and *Phyllanthus amarus.*

S. No.	Features	*Phyllanthus fraternus*	*Phyllanthus amarus*
1	Shape	Oblong	Elliptic-oblong
2	Apex	Obtuse	Obtuse
3	Base	Rounded	Rounded
4	Margin	Entire	Entire
5	Petiole	Short	Very short
6	Colour	Greenish brown	Dark green
7	Odour	Indistinct	Characteristic

Powdered Crude Drug Microscopy of Leaves and Barks. https://doi.org/10.1016/B978-0-12-818092-1.00055-5

FIG. 7.11.1 *Phyllanthus fraternus* leaf.

FIG. 7.11.2 *Phyllanthus amarus* leaf.

7.11.1 Powder microscopic description

P. fraternus leaf powder is greenish brown in colour showing fragments of upper and lower epidermis with wavy anticlinical walls embedded with paracytic stomata; Few rosette and cluster crystals of calcium oxalate; Multicellular covering trichomes with oval head and longitudinally cut spiral xylem vessels.

P. amarus leaf powder is dark green in colour showing simple, unicellular, covering trichomes with pointed apex; Few rosette crystals of calcium oxalate; Fragments of longitudinally cut spiral and annular vessels. Stomata is not observed in leaf powder.

Based upon the powder microscopic features, the leaf powders of *P. fraternus* and *P. amarus* can be distinguished as shown in Table 7.11.2 and Figs. 7.11.3–7.11.5. *P. fraternus* leaf powder shows paracytic stomata whereas *P. amarus* is devoid of stomata. Trichomes are abundant, multicellular with oval head in *P. fraternus* as compared to few, unicellular with pointed apex trichomes in *P. amarus.*

TABLE 7.11.2 Specific comparative feature of powdered leaf powders of *Phyllanthus fraternus* and *Phyllanthus amarus.*

S. No.	Features	*Phyllanthus fraternus*	*Phyllanthus amarus*
1	**Common name**	Bhui-amla	Jar-amla
2	**Stomata**	Paracytic	NA
3	**Trichomes**	Abundant	Few
		Covering	Simple
		Multicellular	Covering
		Warty	Unicellular
		Thick walled	Pointed apex
		Oval head	

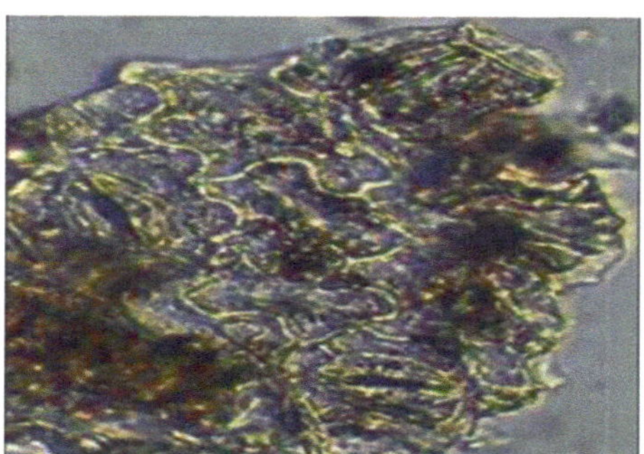

FIG. 7.11.3 Paracytic stomata present in *Phyllanthus fraternus* leaf powder.

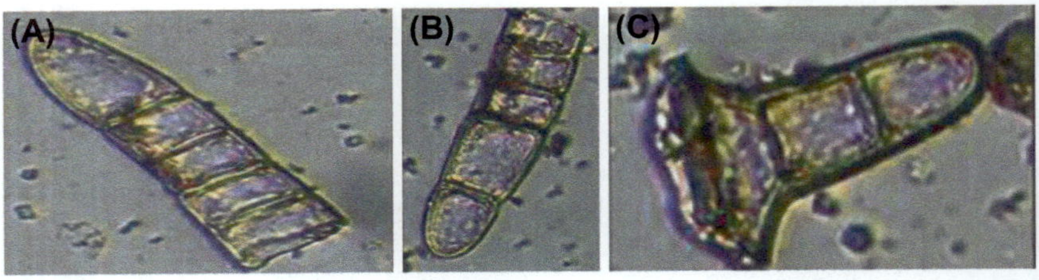

FIG. 7.11.4 (A–C) – Covering trichome present in *Phyllanthus fraternus* leaf powder.

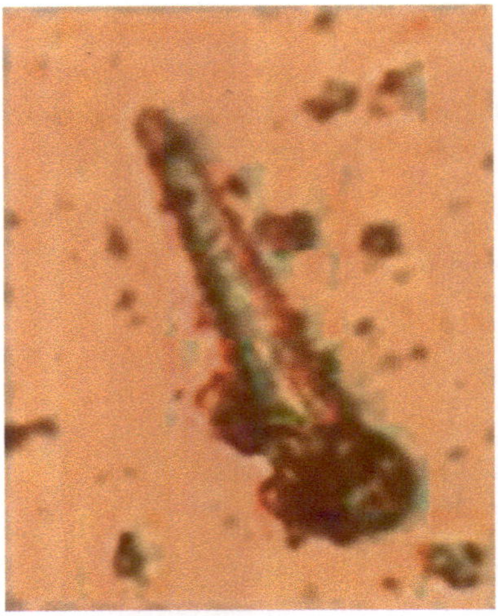

FIG. 7.11.5 Covering trichome present in *Phyllanthus amarus* leaf powder.

Reference

[1] Government of India, Department of AYUSH, Ayurvedic Pharmacopoeia of India. Part I, vol. I, Ministry of Health and Welfare, Ghaziabad, 1999–2017, p. 149.

Chapter 7.12

Vitex negundo and *Vitex trifolia*

The genus *Vitex* contains about 270 known species. *Vitex negundo* and *V. trifolia* are two species commonly found in India and are often substituted with each other.

V. negundo L.	Verbenaceae
Vitex trifolia L.	Verbenaceae

Common vernacular names: Assamese – Aslak; Bengali – Nirgundi, Nishinda; English – Five Leaved Chaste tree; Gujarati – Nagod; Hindi – Nirgundi, Sinduar, Sambhalu; Kannada – Lakkigida, Nekkigida; Malayalam – Indranee, Nirgundi; Marathi – Nirgundi; Punjabi – Sambhalu, Banna; Tamil – Karunochchi, Nocchi; Telugu – Nallavavilli, Vavili [1].

Major chemical constituents: Friedelin, Vitamin C, alkaloids and essential oils [2].

Traditional uses: Ear pain, obesity, diabetes [1].

Macroscopic description: The morphological characters of *V. negundo* and *V. trifolia* are illustrated in Table 7.12.1 and Figs. 7.12.1 and 7.12.2 respectively.

7.12.1 Powder microscopic description

V. negundo is a dark green colour powder showing plenty of simple, covering, uniseriate, multicellular trichomes or their fragments scattered as such or attached to the cells of the epidermis, majority of them being two to four celled, straight, bent or curved with pointed apex and with or without warts; Fragments of upper and lower epidermis with almost straight anticlinical walls and plenty of simple covering trichomes and cicatrix, embedded with anomocytic and paracytic stomata; Glandular trichomes are few, short, with unicellular stalk and bicellular head; Cluster crystals of calcium oxalate and longitudinally cut fragments of annular vessels [3].

V. trifolia is a light green colour powder showing fragments of upper and lower epidermis with straight and anticlinical walls devoid of stomata; Plenty of simple, covering trichomes of various shapes and sizes, majority of them being multicellular, uniseriate, thick or thin walled, short, straight and warty; Few sessile glandular trichomes and prismatic crystals of calcium oxalate.

Based upon the powder microscopic features, the leaf powders of *V. negundo* and *V. trifolia* can be distinguished as shown in Table 7.12.2 and

Powdered Crude Drug Microscopy of Leaves and Barks. https://doi.org/10.1016/B978-0-12-818092-1.00056-7

TABLE 7.12.1 Morphological characters of *Vitex negundo* and *Vitex trifolia*.

S. No.	Features	*Vitex negundo*	*Vitex trifolia*
1	Shape	Leaflet lanceolate	Oblong-elliptic to Obovate
2	Apex	Acute	Acute to acuminate
3	Base	Symmetrical	Cuneate to rounded
4	Margin	Entire	Entire
5	Venation	Pinnate	Pinnate
6	Petiole		Long
6	Petiole	Short	Long
7	Colour	Dark green	Dark green on upper and light green on lower surface
8	Odour	Not characteristic	Characteristic

FIG. 7.12.1 *Vitex negundo* leaf.

Figs. 7.12.3–7.12.7. Leaf powder of *V. negundo* shows anomocytic and paracytic stomata whereas stomata is not observed in *V. trifolia* leaf powder. Numerous simple, covering and glandular trichomes are observed in *V. negundo* as compared sessile sickle shaped trichomes in *V. trifolia*.

FIG. 7.12.2 *Vitex trifolia* leaf.

TABLE 7.12.2 Specific comparative feature of powdered leaf powders of *Vitex negundo* and *Vitex trifolia*.

S. No.	Features	*Vitex negundo*	*Vitex trifolia*
1	**Common name**	Nirgundi	Pani-ki-Sanbhalu
2	**Stomata**	Anomocytic	Devoid of stomata
		Paracytic	
3	**Trichomes**	Abundant	Abundant
		Simple	Simple
		Covering	Covering
		Uniseriate	Multicellular
		Multicellular	Uniseriate
		Straight	Thick walled
		Bent	Short
		Curved	Straight
		Pointed apex	Wavy
		Glandular	Sessile
		Globular head	Sickle shaped
		Bicellular head	Warty
		Hook shaped	Glandular
		Long	Long
4	**Calcium oxalate crystals**	Cluster	Prismatic

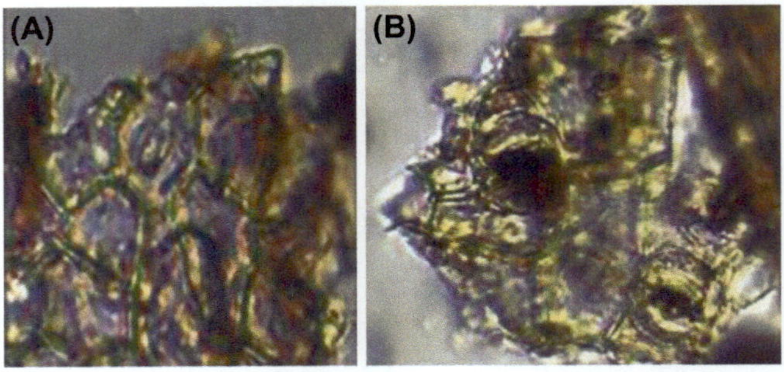

FIG. 7.12.3 (A–B) – Anomocytic and paracytic stomata present in *Vitex negundo* leaf powder.

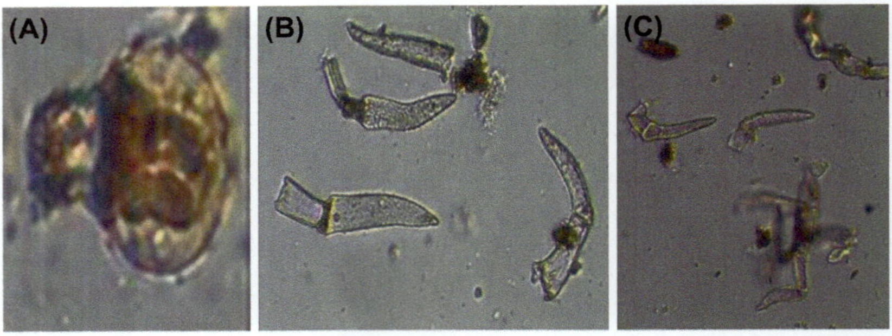

FIG. 7.12.4 (A–C) – Different types of trichomes of various shapes and sizes present in *Vitex negundo* leaf powder.

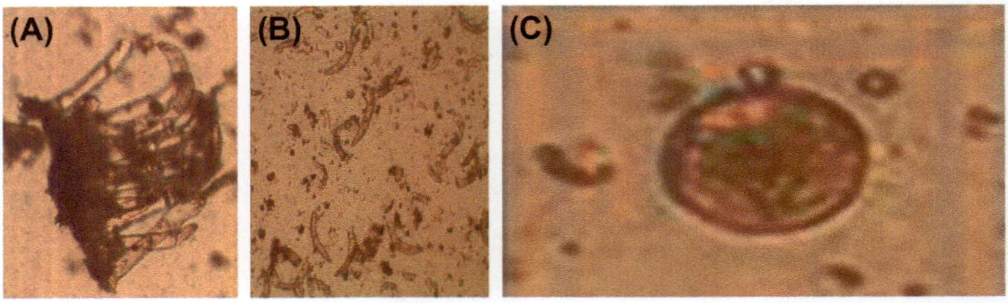

FIG. 7.12.5 (A–C) – Different types of covering trichomes present in *Vitex trifolia* leaf powder.

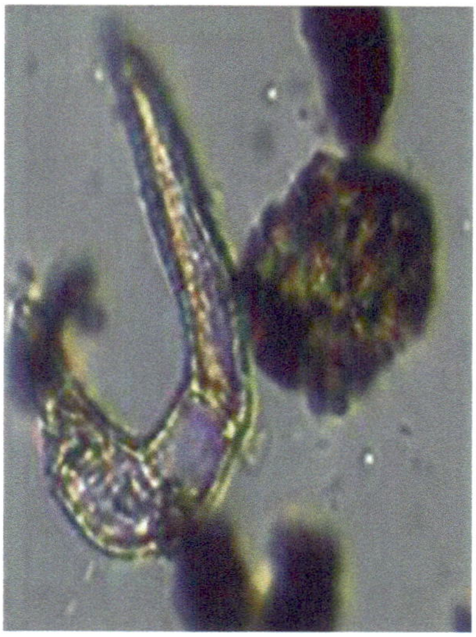

FIG. 7.12.6 Rosette crystals present in *Vitex negundo* leaf powder.

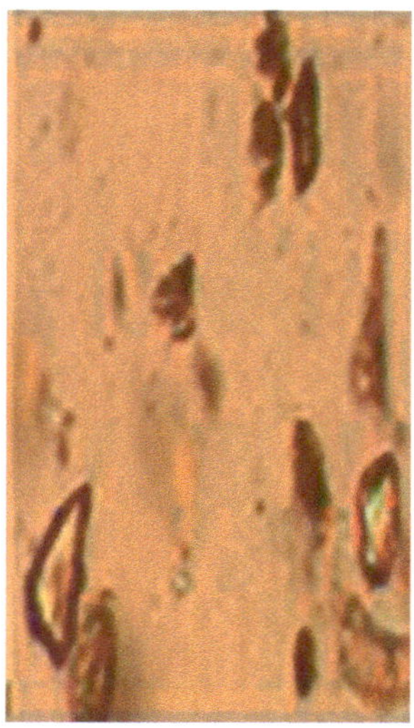

FIG. 7.12.7 Prismatic crystals present in *Vitex trifolia* leaf powder.

References

[1] Government of India, Department of AYUSH, Ayurvedic Pharmacopoeia of India. Part I, vol. III, Ministry of Health and Welfare, Ghaziabad, 1999–2017, p. 144.

[2] A.K. Gupta, N. Tandon, M. Sharma, Quality Standards of Indian Medicinal Plants, vol. 3, Medicinal Plant Unit, ICMR, New Delhi, India, 2008, p. 356.

[3] M.G. Chauhan, A.P.G. Pillai, Microscopic Profile of Powdered Drugs Used in Indian System of Medicine, Leaf Drugs, vol. 2, Institute of Ayurvedic Medicinal Plant Sciences, Jamnagar, Gujarat, India, 2007, p. 100.

Appendix

Mountants, clearing and histochemical reagents

Crude powdered drugs in addition to being used as a mode of studying the microscopic character are also used effectively to study the chemical constituents present in the drug. Application of physical and chemical tests on milligram quantities of the drug powder or using micro quantities of reagents and observing the color/precipitate or crystals is known as microchemistry.

Mountants

Glycerin solution: Aqueous solution of glycerin (50%) used as a temporary mountant.

Lactophenol: Mountant for non-oily powders, starch and pollen grains; the solution is composed of 20 g of phenol, 20 g of lactic acid and 40 g of glycerin in 20 mL of distilled water.

Clearing reagents

Acetic acid: Distinguish calcium carbonate (soluble) from calcium oxalate (insoluble).

Chloral hydrate: Dissolve cell contents like chlorophyll and coloring matter; gelatinizes the starch on boiling and thereby increases the visibility of cell contents. The solution is prepared by dissolving five parts by weight of chloral hydrate in two parts by volume of distilled water.

Hydrochloric acid: Removes starch, dissolves calcium carbonate and calcium oxalate.

Hydrofluric acid: Softens hard tissues of the plant without destroying the structure of cell wall. Generally used to destroy the swelling property of mucilage and makes the mucilage containing drugs easy to handle.

Sodium or potassium hydroxide: Solution (2%) clears and disintegrates non lignified elements.

Potassium chlorate: Potassium chlorate (oxidizing agent) and 50% nitric acid forms a liquid known as Schulze's maceration fluid which is used to disintegrate hard woody tissues of a bark powder such as sclereids and stone cells.

Histochemical reagents

Ferric Chloride: Solution (5%w/v in water) detects the presence of tannins. The tissue containing tannins turns bluish to grayish black.

Iodine solution N/10: Aqueous solution of iodine (12.6 g) and potassium iodide (18 g) in 1000 mL of distilled water imparts yellow color to aleurone grains, latex granules and plamodesma; blue color to starch and brown color to amylodextrin.

Ruthenium red: Solution (0.008 g of ruthenium in 10 mL of 10% led acetate) imparts pink color to mucilage.

Phloroglucinol hydrochloric acid: Solution phloroglucinol 1% in alcohol 90% stains lignified elements followed by addition of hydrochloric acid imparts pink or red color.

Sudan III: Solution (0.5 g in 100 mL of glacial acetic acid) stains oil globules pink.

Sodium or potassium hydroxide: Solution (20%) imparts yellow color to suberized wall, on warming dissolves suberin with formation of yellow droplets of soap. Free anthraquinone develops pink color with potassium hydroxide solution.

Bibliography

American Herbal Pharmacopoeia, Roy Upton Herbalist, 1999.

Ayurvedic Formulary of India, Ministry of Health and Family Welfare, second ed., vol. Part I, Government of India, 2003, pp. 103–119.

British Pharmacopoeia Commission, British Pharmacopoeia, The Stationery Office on Behalf of the Medicines and Healthcare Products Regulatory Agency, London, 2008.

M.G. Chauhan, A.P.G. Pillai, Microscopic Profile of Powdered Drugs Used in Indian System of Medicine, Bark Drugs, vols. 1–2, Institute of Ayurvedic Medicinal Plant Sciences, Jamnagar, Gujarat, India, 2005.

European Medicine Evaluation Agency, Guidelines on Quality of Herbal Medicinal Products, 2006. EMEA/CVMP/814/00. London.

European Pharmacopoeia Commission, European Pharmacopoeia, sixth ed., European Directorate for the Quality of Medicines and Healthcare of Council of Europe, Strasbourg, 2007.

K.M. Ganeshaiah, R. Ganeshan, R.U. Shankar, C. Meera, A taxonomic hurdle by taxonomists, AMRUTH (1998) 3–8.

Government of India, Department of AYUSH, Ayurvedic Pharmacopoeia of India. Part I, Vols.. (I–VI), Ministry of Health and welfare, Ghaziabad, 1999–2017.

A.K. Gupta, N. Tandon, M. Sharma, Quality Standards of Indian Medicinal Plants, Vols. (1–11), Medicinal plant unit, ICMR, New Delhi, India, 2008.

Japanese Pharmacopoeia Committee, The Japanese Pharmacopoeia, fifteenth ed., Society of Japanese Pharmacopoeia, Tokyo, 2006.

D.B.A. Narayana, Pharmacognosy and technology, Pharmacognosy Magzine 6 (23) (Jul.–Sep. 2010) 145–146.

M.D. Rajendra, D.B. Ananthanarayana, Controversial nomenclature of ayurvedic drugs. Challenges for scientists, Pharmacognosy Review 3 (5) (2009) 1–7.

R. Reddy, B. Kalpana Vignanam, Chaukhambha Sanskrit Bhawan, fourth ed., 2015, pp. 41–42. Varanasi, India.

D. Singh, V. Aeri, D.B. Ananthanarayana, Development of standard operating protocol for slide preparation of powdered bark samples with varying grinding techniques, Pharmacognosy Journal 10 (2) (2018) 265–271.

V.V. Sivarajan, I. Balchandran, Ayurvedic Drugs and Their Plant Sources, Oxford & IBH Publishing CO. Pvt. Ltd., New Delhi, 1994.

The Committee of the Ayurvedic Pharmacopoeia of India. The Ayurvedic Pharmacopoeia of India. New Delhi: The Controller of Publications Civil Lines.

The Editorial Committee, Materia Medica of China, Shanghai Science andTechnology Publishing House, Shanghai, 1999.

The United States Pharmacopeial Convention, The United States Pharmacopeia, 30th Revision/ National Formulary, twentyfifth ed., The United States Pharmacopeial Convention, Rochville, 2005.

B.L. Vaidya, Some Controversial Drugs in Indian Medicine, Chaukhambha Orientalia, Varanasi, 1982.

Vietnamese Pharmacopoeia Commission, Ministry of Health. Vietnamese Pharmacopoeia, third ed., Ministry of Health – S.R. Vietnam, Hanoi, Vietnam, 2005.

P.K. Warrier, V.P.K. Nambiar, C. Ramankutty (Eds.), Indian Medicinal Plants, vols. 1–5, Orient Black Swan/Universities press, 1994.

World Health Organisation, Quality Control Methods for Medicinal Plants, 1998. Geneva.

World Health Organisation, Traditional Medicine Strategy, 2002–2005. Geneva.

World Health Organisation, Guidelines on Good Agricultural and Collection Practices (GACP) for Medicinal Plants, 2003. Geneva.

Botanical name index

Names of other languages

C

D

E

F

G

L

M

N

P

R

S

T

U

V

Y

Common name index

'*Note*: Page numbers followed by "f" indicate figures and "t" indicates tables.'

FSC
www.fsc.org
MIX
Papier | Fördert
gute Waldnutzung
FSC® C083411

Kolibri 360 GmbH
Ferdinand-Jühlke-Straße 7
99095 Erfurt, Deutschland
produktsicherheit@kolibri360.de